12.50

FLOWERING PLANTS
ORIGIN AND DISPERSAL

FLOWERING PLANTS

ORIGIN AND DISPERSAL

ARMEN TAKHTAJAN
D.Sc., Corr. Memb. Acad. Sc. U.S.S.R., F.M.L.S.
Department of Higher Plants, Botanical Institute of the Academy of Sciences of the U.S.S.R., Leningrad

Authorised translation from the Russian by

C. JEFFREY
B.A.
Senior Scientific Officer, Royal Botanic Gardens, Kew

OLIVER & BOYD : EDINBURGH

ENGLISH EDITION
First published 1969

This is a translation of Порисхождение покрытосеменных растений, Изд. 2-е (The Origin of Angiospermous Plants. 2nd ed.) published in Moscow in 1961, which has been extensively revised and to which has been added much new and previously unpublished material.

OLIVER AND BOYD LTD
Tweeddale Court, Edinburgh 1

05 001715 2

Printed in Great Britain by
Robert Cunningham and Sons Ltd, Alva

PREFACE

It has given me great pleasure to accept Oliver and Boyd's proposal that I should acquaint the English reader with my work on the origin and dispersal of flowering plants. The original intention of the publishers was to provide a translation of my book *The Origin of Angiospermous Plants*, the second edition of which was published in Moscow in 1961. During its preparation for translation, however, and also after translation, the manuscript (and even the galley) has been subjected to so much revision that it is now substantially different from the Russian edition. It is in fact a new book that is here presented to the reader, and for this reason I have permitted myself to accord it a new title.

The origin and dispersal of the flowering plants is one of the most troublesome and complex questions of biological history, touching as it does upon many of the general problems of organic evolution. Much is uncertain, and much is in dispute. There are so many different hypotheses of the origin of flowering plants, and so many different attempts at morphological interpretation of the flower, that even a brief résumé of them, let alone a critical review, would require a whole volume. I am therefore unable to allow myself a discussion of the numerous theories of the flower that have appeared from time to time in the botanical literature, and indeed it is not my aim to do so. I have confined myself to stating, and as far as possible substantiating, those ideas and concepts that seem to me to be correct or at least plausible, and in only a few cases have I mentioned the opposing points of view. Among these few exceptions are a critique of the idea of a polyphyletic origin for the flowering plants and a critique of the hypothesis of their origin in high latitudes. As far as the various theories of the flower which dispute the sporophyll nature of the stamens and carpels are concerned, the reader will find a critical analysis of them in the papers of John Parkin and Professor Arthur Eames and also in the latter's *Morphology of the Angiosperms* (1961).

The number of works devoted to some or other aspect of the problem of the origin and evolution of the flower and of flowering plants is very great and is rapidly increasing. Therefore in the writing of this book some subjectivity was inevitable in the choice of questions to be discussed and in the choice of material to be presented. Naturally I have given greatest attention to those aspects of the problem in which I myself am particularly interested. This book therefore cannot claim to be a full and thorough survey of the present position with respect to the problem of the origin of flowering plants, but if the reader finds his interest aroused by any of the questions raised herein and is stimulated to gain a wider acquaintance with the literature, then the author will consider his aim to have been fulfilled. The reader who desires a fuller acquaintance with various other contemporary views on the origin of the flower and the evolution of flowering plants should consult the works of Croizat (1960), Lam (1961), Melville (1962, 1963), and Meeuse (1965).

I cannot conclude this preface without expressing my sincere gratitude to all who have helped in the preparation of this book—those who have provided me with invaluable data, with illustrations, and with material for illustrations. There are many. But I thank in the first place Mr C. Jeffrey, who kindly undertook the onerous task of translation and has provided the book with an index. The translation was carried out so well that I had to make only a few minor corrections. In addition, Mr Jeffrey made a number of valuable suggestions, which were very helpful during the author's revision of the translated manuscript. Finally, Mr Jeffrey has given me considerable help in the selection of illustrations.

For their patience and for the unfailing attention that they have given to the publication of this book, I accord my gratitude to Oliver and Boyd.

A. Takhtajan

Leningrad

TRANSLATOR'S PREFACE

It is a great privilege to have been given the opportunity of bringing Professor Takhtajan's book to the notice of a wider public through the medium of the English language. Professor Takhtajan, Chief of the Department of Higher Plants of the Komarov Botanical Institute of the Academy of Sciences of the U.S.S.R., in Leningrad, is one of the leading Soviet plant taxonomists and morphologists of today and an authority of world repute on the classification, origin and historical phytogeography of the flowering plants.

The past few years have been marked by an upsurge of interest in the origin of the flowering plants, and as a result several new theories have been published concerning their morphology and derivation. Unfortunately, this growth of theories has not been paralleled by a commensurate increase in our knowledge of the earliest angiosperms. In a field where facts are few and speculation is rife, student and specialist alike will surely welcome Professor Takhtajan's book, which gives such facts as are known and from them draws the conclusions that appear to the author to be most probable. The very full bibliography of both Western and Russian literature will likewise be most welcome.

In conclusion, I wish to thank several people who have helped in various ways. Those who have kindly lent photographs to help with the illustrations will find their contributions duly acknowledged in the appropriate places. Above all, I thank Professor Takhtajan for his patience, co-operation, advice and constant encouragement. To Mrs Angela Barnwell, who typed the English manuscript, and Mrs Ruth Jeffrey, who typed the index, I wish to express my gratitude. Finally, I am greatly indebted to Dr A. Melderis for providing the basis of the translation of Chapters 1 and 2.

C. JEFFREY

Ex quo exstitit illud, multa esse probabilia, quae quanquam non perciperentur, tamen, quia visum quendam haberent insignem et illustrem, iis sapientis vita regeretur.

CICERO, *De Natura Deorum*, i, 5

CONTENTS

CHAPTER 1

INTRODUCTION

The flowering plants, or angiosperms (*Magnoliophyta* or *Angiospermae*), are the largest and most diversified group in the plant world today. As land plants they predominate over all other groups, not only in the number of their species—at least 250 000*—but also in their enormous number of individuals. In most places the vegetative cover of the land is for the greater part angiospermous. Only in areas of moss and lichen tundra, in fern and cycad thickets and in the large areas of coniferous forest are the flowering plants outnumbered by other plants. Even in these areas, however, they often play a fairly conspicuous role, and in coniferous forests they may predominate in terms of the number of species and individuals.

A feature of the flowering plants is their tremendous evolutionary plasticity, which far exceeds that of the ferns and gymnosperms. In the course of evolution they have become adapted to the most diverse environmental conditions and may be found wherever higher plants can survive—from the frigid rocks of the highest Himalayas (Plate III) to the torrid, salty and sandy deserts of Asia (Plate II) and Africa. They play an important role also in fresh waters; and have even become adapted to a marine existence, although here they are represented by only a few dozen species that grow in shallow water on sandy shores. As Scott (1911) remarked, 'The angiosperms alone, of all the higher groups of plants, appear to have any effect at all on the flora of the sea' (Plate IV). Forms with sufficient plasticity to become adapted to a marine existence are not found amongst the mosses, hepatics, club-mosses, horsetails, ferns and gymnosperms. The range of angiosperm adaptability is thus extremely great.

The flowering plants have evolved a remarkable variety of life-forms (Plates V, VI), ranging from such minute plants as the

* However, the angiosperms take first place only in the plant world; the number of insect species is several times greater.

duckweed *Wolffia* to gigantic tropical forest trees like *Eucalyptus regnans*, which may reach a height of nearly 100 metres (Scott, 1911). Besides typical green photosynthetic plants there are numerous saprophytes and parasites. Especially characteristic of tropical and subtropical regions are various types of lianas which, as creepers, twiners or climbers, employ a variety of methods of attachment to their supports. The climbers are especially diverse in this respect; Krassnov (1956) vividly describes how every organ of the plant body may be employed by such plants in their ascent to the sunlight. In the tropical rain forests (Plate I) many angiosperms have become adapted to life on the trunks, the branches and even the leaves of the various woody plants—these are the epiphytes, which are not parasitic but merely enjoy the anchorage afforded by other plants.

The epiphytes have developed many peculiar and distinctive features as adaptations to their unusual mode of life. A remarkable example is the tropical *Dischidia rafflesiana* (Fig. 1), which has not only ordinary leaves but also modified, flask-like leaves that collect rain-water and soil particles carried in by the ants which inhabit them. In these 'natural flower-pots' (as Scott called them) highly-branched accessory roots develop which absorb water and nutritive substances. There are numerous epiphytes amongst the ferns, but they seldom show such complicated adaptations. Even more astonishing adaptations may be seen in the various carnivorous angiosperms, usually referred to as 'insectivorous' plants, although they do not feed exclusively on insects. There are no carnivorous monocots; but amongst the dicots carnivores are found in the families *Cephalotaceae*, *Droseraceae*, *Lentibulariaceae*, *Nepenthaceae* and *Sarraceniaceae* (see Skene, 1948). Despite their small number these genera display a diversity of adaptations for trapping small animals. What wiles these predators employ to catch their victims! The Sundew (*Drosera*) traps small creatures such as insects with its special viscid glands; in Venus's Flytrap (*Dionaea*, Plate VIII) the leaf-lobes suddenly snap together and the prey is actively caught; in others, such as the tropical pitcher-plants (*Nepenthes*, Plate VII), the victims may be caught in special flask-like structures or (as in the widespread bladderworts of the genus *Utricularia*) in tiny bladder-like traps, and their remains absorbed. The minute bladders of *Utricularia* are of complex and ingenious construction; their mechanism was explained by Lloyd (1942).

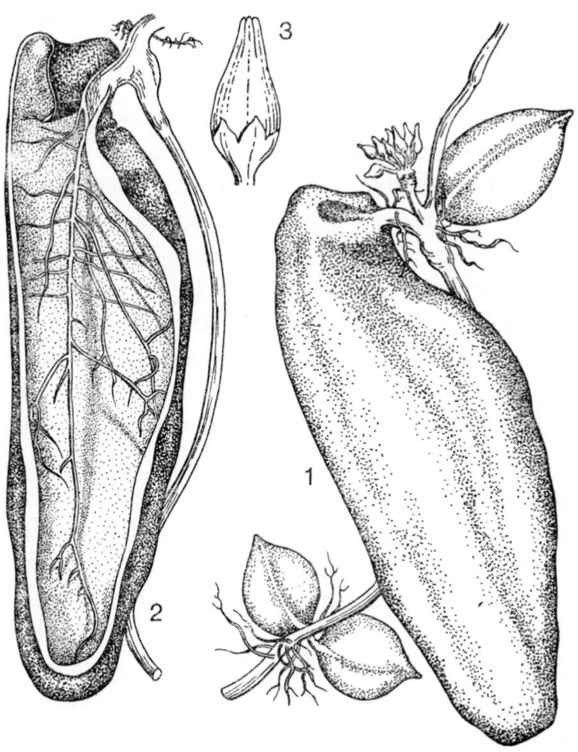

Fig. 1. *Dischidia rafflesiana*: *1*, shoot with ordinary leaves and one pitcher-leaf; *2*, pitcher-leaf in longitudinal section, showing adventitious roots; *3*, flower-bud. (After Wettstein, 1935.)

Among the flowering plants there are also various types of parasites (Plates IX, X) which live at the expense of other plants by absorbing nutrients from them. Well-known examples are the Broomrape (*Orobanche*) and the Dodder (*Cuscuta*). The famous parasite *Rafflesia arnoldii* (Plate IX) of Sumatra is of especial interest. Its vegetative organs have become highly modified and simplified and serve only as a system for absorbing nutritive substances from the roots of the host. Its flower, on the other hand, has attained enormous dimensions and measures over 1 m across. The most varied adaptations are found in plants of the deserts and fresh waters. As for the adaptations associated with flower-pollination and dispersal of fruits and seeds, their diversity exceeds all imagination—the inexhaustible creativity of nature here appears in its fullest measure. With their great variety of form and function, the angiosperms occupy a place in the plant world comparable with that held by the mammals amongst the animals.

The asexual generation, or sporophyte, which forms the plant as we know it, has become exceedingly complex—both externally, and internally in its histological differentiation. The leaves show an extraordinary plasticity, the vascular tissues the highest degree of elaboration, and the mechanical, dermal and storage tissues the greatest efficiency and perfection. The angiosperm flower is much more complicated and diversified than the gymnosperm cone, or strobilus. In contrast, the sexual generation is greatly simplified and reduced: both male and female gametophytes are formed and function in the swiftest and most economical manner, requiring the minimum number of cell divisions; antheridia and archegonia—the male and female sex organs—are not formed, and the gametophytes therefore utilise the minimum amounts of material and energy. A peculiar 'double fertilisation' has evolved in the flowering plants, which gives them a great biological advantage over the gymnosperms.

The morphological elaboration of the flowering plants is closely associated with their outstanding functional efficiency; they show a great complexity in their physiological division of labour and the highest degree of efficiency in the performance of the functions themselves. From his studies of the effect of light on the development of pigmentation, photosynthesis and the accumulation of organic substances, Lubimenko (1933) concluded that adaptation to phototrophic existence has developed side by side with the evolution of plant form, and that the angiosperms appear in this respect to have attained the highest degree of adaptation and a great potentiality for the evolution of heliophilic forms. They are distinguished by a high degree of light-sensitivity, which has led to the development of such special reactions as the photoperiodism which is so well exhibited by annual herbs.

The time of the first appearance of the flowering plants, the place of their origin, their ancestral gymnosperms, their main lines of evolution—these are some of the questions with which we shall now be concerned.

CHAPTER 2

THE FLOWERING PLANTS—ARE THEY POLYPHYLETIC?

Having undergone intensive evolution in many directions under many different environmental conditions, the flowering plants have attained such an extraordinary degree of morphological and ecological diversity that it is sometimes suggested that they must be 'polyphyletic' in origin—in other words, must be derived from more than one ancestral source.

The well-known palaeozoologist Abel (1929) pointed out that the idea of a 'polyphyletic origin' for any group should be eradicated from science because of its incongruity. Thus, if further research showed that a systematic group formerly considered monophyletic (i.e. natural) was composed of elements of various origins, then that group would have to be broken up into as many taxonomic units as there were phylogenetic lines within it. Should, for example, the monocots and dicots be shown to have originated from different ancestral groups, they could no longer be logically left as forming the single taxon—the flowering plants, or angiosperms.

In fact, the morphological and taxonomic evidence leads to the conclusion that the flowering plants are one natural monophyletic branch of development and are not made up of various unrelated parallel branches of different origins. The common origin* of all the orders and families of angiosperms is demonstrated above all by their multitude of common morphological characters. It is demonstrated by the uniform staminal structure, with the characteristic endothecial layer of the anther-wall, by the presence of specialised megasporophylls (carpels) with stigmas, by the constancy of the

* 'Common origin' is here understood to mean origin not simply from one pair of individuals but (for a species) from a population or (for taxa of higher rank) at least from a single taxon of lower rank. (See also Simpson, 1961; Remane, 1956; Heslop-Harrison, 1958; Davis and Heywood, 1963; Tikhomirov, 1965.)

relative position of the androecium and gynoecium on the floral axis, by the characteristic male and female gametophytes, basically similar throughout the group and accompanied by double fertilisation and the formation of a triploid endosperm, and by the presence of sieve tubes. As Parkin (1923) pointed out, the independent origin of these characters in different taxonomic groups is statistically very unlikely. He wrote (p. 56), 'Even admitting the possibility of the angiospermous embryo-sac as having arisen independently more than once, the chance of its being associated each time with the same kind of microsporophyll would be extremely unlikely.'* (See also Eames 1961, p. 467.) The force of this argument against the polyphyletic hypothesis has recently been re-emphasised by Yatsenko-Khmelevsky (1957), in whose opinion the fact that the angiosperms are characterised by a whole series of unique characters which are independent of one another in their development is the basic argument in favour of their being monophyletic in origin. He writes: 'It is quite improbable, and mathematically demonstrable as improbable, that two (or more) groups of different origin could develop such similar sets of characters, although these characters are uncorrelated in themselves.' The independent origin of the whole complex of angiosperm characters in several different lines of higher plants (or even gymnosperms alone) would be comparable to the so-called 'dactylographic miracle' whereby given enough time, a work of literature, such as a Shakespearian sonnet, would be produced by chance alone.

However, the naturalness and common origin of the flowering plants are proved, not only by their possession of a great number of independent common characters (e.g. structure of the female gametophyte and of the stamens) and by the mathematical (probability theory) demonstration of the infinitely small probability of

* It should be pointed out, however, that the independent origin of the angiosperm embryo-sac in different lines of development is very improbable if one bears in mind such distinctive structures and features of the female gametophyte as its marked polarity (with characteristic synergid and antipodal regions), the fusion of the two polar nuclei to form the diploid nucleus of the central cell (which later fuses with one of the two male nuclei to form the endosperm), and so on. The statistical probability of the independent origin in various lines of evolution of this whole complex system is extremely small.

their appearance together in one taxonomic unit (given their absence from all other taxa), but also by direct taxonomic evidence. Thus, for example, the orders *Urticales*, *Casuarinales* (Fig. 26/*8–12*, p. 103) and *Fagales*, so distinct at first sight from the *Magnoliales* (Fig. 14, p. 75), are linked with them by an intermediate group, the *Hamamelidales* (Figs. 23, 24, pp. 99, 100); and the *Lemnaceae*, which appear to resemble the *Araceae* even less than the *Urticales* resemble the *Magnoliales*, are in fact closely connected with them through the somewhat intermediate genus *Pistia*. One may instance numerous such examples, and they show that those groups which at first sight seem completely isolated and suggest a polyphyletic origin do, after closer investigation and wider comparison with other groups, in fact sooner or later fall into their natural place in the system of angiosperm classification.

CHAPTER 3

ANCESTORS UNKNOWN?

The identity of the ancestors of the flowering plants is a most difficult problem which is as yet far from being solved. It would be hard to find a group of plants that has not at some time been postulated as ancestral to the flowering plants; not only the various gymnosperm groups, such as the pteridosperms (including the *Caytoniales*), cycads, *Bennettitales*, *Cordaitales*, conifers, *Gnetum*, and *Ephedra*, but even ferns, equisetophytes, lycopodiophytes, rhyniophytes ('psilophytes')* and algae have been considered as possible precursors (see Tikhomirov, 1965). Yet these very diverse hypotheses have on the whole been given little justification by their authors; indeed, some have not even bothered to accord their views any factual basis whatsoever. Most have not withstood the test of time and have been discarded, but some do continue to command support. Can we find which in fact comes closest to the truth?

The question will be tackled by a process of elimination. First, we must accept that there is no factual basis of any kind for attempts to derive the angiosperms directly from ferns or other lower *Embryobionta* without a transitional gymnospermous stage. All available evidence indicates that the seed-plants, from pteridosperms to angiosperms, have arisen from only one ancestral fern-like group. In spite of their great diversity, all seed-plants have so much in common that their origin from more than one ancestral group seems unlikely.† In this respect the ovule is most important; it has a

* This book follows the system and the nomenclature of the higher taxa of *Embryobionta* (*Cormophyta*) proposed in Cronquist, Takhtajan, and Zimmermann's article 'On the higher taxa of *Embryobionta*' (1966). We divide the subkingdom *Embryobionta* into eight divisions: *Rhyniophyta* (*Psilophyta*), *Bryophyta*, *Psilotophyta*, *Lycopodiophyta*, *Equisetophyta*, *Polypodiophyta* (ferns), *Pinophyta* (gymnosperms) and *Magnoliophyta* (angiosperms). *Pinophyta* are divided in their turn into subdivisions *Cycadicae*, *Pinicae* and *Gneticae*, and *Magnoliophyta* are divided into classes *Magnoliatae* (*Dicotyledones*) and *Liliatae* (*Monocotyledones*).

† According to Beck (1966), 'the cycadophytic and coniferophytic gymno-

common plan of construction in all seed-plants, and the question of its origin merits further consideration.

The ovule is a distinctive and highly diversified structure, but in all groups of seed-plants it consists of the same two basic parts—megasporangium (nucellus) and surrounding protective coverings (integuments). In the nucellus are found the two essential elements of any sporangium, the sporogenous tissue (usually much reduced) and the wall (albeit greatly simplified histologically). It is clear that the angiosperm nucellus is fully equivalent to that of the gymnosperms, i.e. they are homologous structures, and its morphological nature as a megasporangium (established by Hoffmeister in the middle of the last century) has rarely been doubted. The interpretation of the integuments is more difficult. Their nature was for long a mystery until Margaret Benson (1904) proposed a very interesting and plausible theory of their origin.

On Benson's hypothesis, the ovule is a synangium in which all but one of the sporangia have become sterile and have taken on the role of a protective covering for the remaining fertile sporangium. In other words, the integument represents the sterile modification of a ring of fused sister-sporangia surrounding the central fertile megasporangium (nucellus), and the micropyle represents the gap between the apices of the original sporangia. According to Benson, each longitudinal segment of the multi-chambered integuments of certain palaeozoic seed-ferns corresponds to one of the sterilised sister-sporangia. This ingenious and plausible hypothesis was accepted by many notable morphologists and palaeobotanists. According to Kozo-Poljanski (1928), for example, the components of the integument are the equivalents of sterile sporangia, the ovule being a synangium of which only the central sporangium (nucellus) is fertile. Halle (1933), Thomas (1936) and Meeuse (1963) also favoured Benson's hypothesis. Teratological conditions of certain conifer ovules have also been used, as by Doyle and O'Leary (1934), in its support. But the best evidence in support of Benson is afforded by the primitive ovules of

sperms might have had a common origin in the *Aneurophytales*, the former evolving by way of calamopityeans, the latter by way of *Archaeopteris* or some similar genus' (p. 338).

the pteridosperms, which in many cases have preserved clear traces of their synangial origin. Recent researches on the ovules of many seed-ferns, especially ones such as *Tyliosperma orbiculatum* with the distal part of the integuments deeply lobed, have fully substantiated the theory of the synangial origin of the ovule. *Physostoma elegans*,

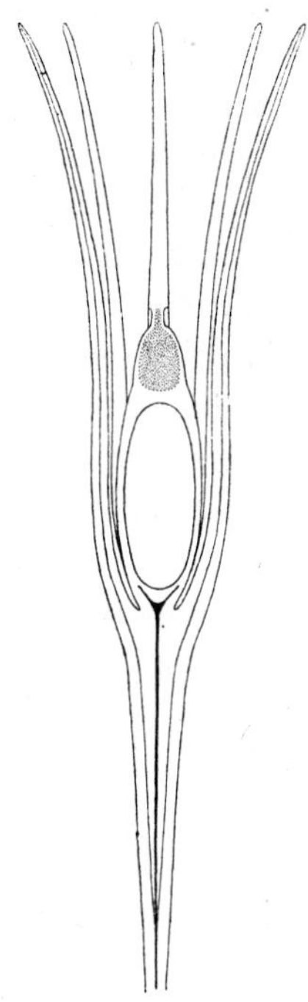

Fig. 2. Diagrammatic longitudinal section of *Genomosperma kidstonii* with the nucellus loosely surrounded by elongated sterile appendages. (From Delevoryas, 1962, after Long, 1960.)

in which the pollen chamber is surrounded by 12 tentacles representing the free end-portions of the sterile sporangia (integument-chambers), is of particular interest in this respect. Especially interesting also is an Early Carboniferous seed-like structure described under the name of *Genomosperma kidstonii* (Long, 1960) in which a whorl of eight elongated processes forms an open cup around the nucellus (Fig. 2); no distinct micropyle is formed. Segmented integuments are also shown by other pteridosperms. It is interesting, however, that they are known not only in seed-ferns but also in several *Bennettitales* and cycads. Thus, the thick integument of *Cycadeoidea morieri* is divided by radial plates of thickened cells into four distinct longitudinal chambers. In *Macrozamia*, *Ceratozamia*, *Encephalartos* and other cycads (de Haan, 1920) the sclerotesta is divided around the micropyle into 7–16 lobes which correspond to the tips of the segments of the integument.

All this supports Benson's theory. More recently, however, under the influence of the telome theory in plant morphology, Benson's hypothesis has undergone some modification. The telome variant of the synangial hypothesis was proposed by some authors. According to Kozo-Poljanski (1948), for example, the morphology of many Silurian plants, such as *Yarravia*, *Hedeia* and others, suggests the possibility of the direct origin of the ovule from a compact 'cluster' of shortened telomes. Florin (1951), Zimmermann (1959, 1965), Delevoryas (1962), Camp and Hubbard (1963), Smith (1964), and others have held more or less similar views. But whether one accepts Benson's original hypothesis or its telomic modification, the ovules of all groups of seed-plants have probably had a common origin and represent variations on a common structural theme. The multiple origin of the ovule seems to me less probable, though quite possible. In his interesting paper on the evolution of the ovule, Smith (1964) comes to the conclusion that 'within the pteridosperms the ovule has arisen independently a number of times' (p. 154), but he adds that the pteridosperms 'may still have been monophyletic in that their pteridophytic ancestors were, perhaps, closely related and had arisen from a common stock'.

We can therefore take the ovule as having first evolved in one particular line of ancient heterosporous ferns or fern-like plants. Thus the seed-plants are most probably connected with the ferns only by way of their most primitive representatives, the seed-ferns. Not a single group of seed-plants is known of which it could be said with certainty that it came directly from the ferns or (even less likely) from the rhyniophytes, and bypassed the intermediate stage of the seed-ferns.* As for the other groups of higher plants, such as the lycopodiophytes and the equisetophytes, it is now considered sufficiently well demonstrated that they represent completely

* But see Cronquist (1960), who argues that 'coniferophytes and cycadophytes do not have a common ancestor short of the psilophytes' (p. 472).

separate and independent evolutionary lines, which likewise had their origin in the rhyniophytes but developed in different directions sharply distinct from the evolutionary line of the ferns and seed-plants. These different branches of the genealogical tree form the different divisions (or phyla) of higher plants. Ovules are found only in gymnosperms and angiosperms, and are unknown in lycopodiophytes* and equisetophytes.

The ancestors of the angiosperms must thus be sought amongst the gymnosperms. But the gymnosperms consist of groups at very different levels of organisation, specialised to varying degrees in their vegetative and reproductive parts. The majority of gymnosperms, especially contemporary ones, have in many respects attained an evolutionary level higher than that of some of the primitive angiosperms. For instance, the living genera *Ephedra*, *Welwitschia* and *Gnetum* have vessels in the secondary wood; yet some angiosperm families, such as the *Winteraceae* (woody) and *Nymphaeaceae* (herbaceous) have xylem completely devoid of them. Furthermore, special investigations (Thompson, 1918) have shown that the vessels of *Welwitschia*, *Ephedra* and *Gnetum* originated in an entirely different way from those of angiosperms. Thus the wood anatomy shows that the angiosperms could not have arisen from any of these three genera.

Some vesselless angiosperms, like *Trochodendron*, *Tetracentron* (Fig. 4, p. 52), some species of *Bubbia* and of *Drimys*, and the genus *Amborella*, have a structurally more primitive secondary xylem than *Ginkgo*, conifers, *Cordaitales*, *Pentoxylaceae*, some cycads and *Bennettitales* and even the seed-ferns. In these angiosperms the early (spring) wood has scalariform tracheids (with scalariform bordered pits), while in *Ginkgo*, the conifers, *Cordaitales* and *Pentoxylaceae*, in

* It is true that in the Carboniferous genera *Lepidocarpon* and *Miadesmia* the megasporangium, with its included female gametophyte, is surrounded by an almost closed indusium-like structure with a narrow apical slit-like opening. But this sheath, although simulating an integument, represents lateral upgrowths of the sporophyll and is thus quite distinct in origin from the true integument of seed-plants. As Andrews (1961) says, 'It is quite apparent that the "integument" here is not homologous with the integument of the seed-plants' (p. 234). Thus the seed-like structure of *Lepidocarpon* is analogous to, but not homologous with, the ovule (see also Delevoryas, 1962).

the majority of cycads (excluding *Stangeria* and *Zamia*), some *Bennettitales* and in the seed-ferns and *Gnetales*, all the tracheids in the secondary wood have more or less circular bordered pits. As it is generally accepted that in the course of evolution circular bordered pits arose from scalariform ones, the secondary xylem of these vessellesss angiosperms has tracheids more primitive in structure than that of most gymnosperms. In the opinion of Bailey (1944a), the scalariform tracheids of primitive angiosperms exclude the possibility of deriving them from the *Gnetales* or other higher gymnosperms.

Comparison of the secondary xylem of primitive angiosperms with that of various groups of gymnosperms shows us that most gymnosperms have a more highly organised secondary xylem (both early and late) than do the vesselless angiosperms of the *Trochodendron* type. Thus they cannot be ancestors of angiosperms. On the grounds of wood anatomy, we can admit as possible ancestors of angiosperms only a few cycads, most of the *Bennettitales* and, of course, the hypothetical ancestors of these groups themselves—probably seed-ferns with a primitive early secondary xylem—which are as yet undiscovered (see p. 32).

The circle of possible angiosperm ancestors is narrowed still further if we consider the reproductive organs and compare the strobili (cones) of various gymnosperm groups with the primitive type of angiosperm flower. Flowers and cones may be, as is well known, either bisexual (ambisporangiate) or unisexual (monosporangiate). In the former, the strobilus, or flower, contains both microsporophylls with microsporangia and megasporophylls bearing ovules; in the latter it consists solely of microsporophylls or of megasporophylls. The strobili of most gymnosperms are unisexual, bisexual ones being found normally only in a few now extinct genera; but the flowers of the majority of angiosperms are bisexual. However, unisexual flowers are not infrequently found in angiosperms, many families in fact being characterised by them. Can we ascertain which type is primitive and which is derived?

Modern studies on the morphology of the flower have shown that unisexual flowers have arisen as a result of reduction of either the microsporophylls (stamens) or the megasporophylls (carpels)—

see, for example, Arber and Parkin (1907), Hallier (1912a), Kozo-Poljanski (1922, 1928), Parkin (1957), Eames (1961), and Takhtajan (1959, 1964). One piece of evidence for the derived nature of unisexual flowers is that many have vestigial organs or rudiments indicative of a former bisexuality. They occur either as rudiments of microsporophylls—staminodes—in female flowers, or as rudiments of megasporophylls—pistillodes—in male flowers. They may be seen, for example, in the flowers of some *Drimys* species, in *Lindera* and *Sargentodoxa*, in some members of the *Monimiaceae*, *Lardizabalaceae* and *Menispermaceae*, in many *Moraceae* and *Urticaceae*, in *Platanus*, several *Juglandaceae*, in *Garrya*, *Freycinetia*, and in many others. It is very unlikely that they are incipient organs, newly and independently evolving in all these different plants; much more probably they are slowly vanishing relics, and so may be regarded as evidence of the bisexual nature of the flowers of their ancestors. In the evolution of many unisexual flowers the rudiments of stamens or carpels have entirely disappeared, but in these cases all phylogenetic researches lead to the conclusion that plants with such flowers have arisen from bisexual ancestors. Plants with unisexual flowers are also sometimes found with bisexual flowers as an atavistic abnormality.* All these lines of evidence show that the basic type of angiosperm flower was bisexual,† and that the unisexual type is derived and secondary. From

* That atavistic variation can be a reversion to an ancestral form may be explicable by ancient genes that determine vanished ancestral features (such as bisexuality) continuing to exist in the genotype, but in such circumstances that their effects are not normally manifest. According to Huxley (1942), atavisms 'are all due to new combinations of old genes'; see also Heslop-Harrison (1952).

† In spite of the firm basis that exists for this view, some contemporary authors still adhere to the old idea of the primitiveness of the unisexual flower. Parkin, in two very interesting papers (1952, 1957), critically examined the views of those botanists who consider the unisexual flower to be primitive. In opposing these views (1952), he asks: 'How do those botanists, who favour the primitive angiospermous flower as being unisexual, derive the bisexual (hermaphrodite) one from it?' It is in fact impossible to give a satisfactory answer to this question unless we admit the so-called 'pseudanthium' hypothesis, according to which the flowers of angiosperms are derived from the 'inflorescences' of gymnosperms of the *Ephedra* or *Gnetum* type. On this supposition, the bisexual flower is the result of the union of male and female strobili; but it requires some very arbitrary assump-

this it follows that we must seek the ancestors of angiosperms amongst those gymnosperms with bisexual strobili.

All extant gymnosperms have unisexual strobili, and, as the work of palaeobotanists has shown, the overwhelming majority of past gymnosperms were likewise unisexual. Bisexual conifers and ginkgos are unknown, and there were no bisexual members amongst those extinct ancestors of the conifers, the *Cordaitales*; finally, there are no known examples of bisexuality in the strobili of cycads. Evidently, in all these groups of gymnosperms the strobili were constantly unisexual from the start. In all probability they arose from unisexual fertile shoots; in other words, the micro- and megasporophylls of their ancestors were found on different shoots. Of completely different derivation, however, are the strobili of *Welwitschia mirabilis* (*W. bainesii*), a gymnosperm with many outstanding features that grows in the stony deserts of Angola and tropical South-West Africa.

In its microstrobili *Welwitschia* has a well-marked rudimentary ovary with an expanded peltate apex. Some authors consider that this rudimentary ovary acts as a nectary; in this it has taken on a new function. Its structure shows that it is not an incipient organ in course of evolution, but a relic or vestige of a functional ovary. The presence of such a rudiment is explicable only on the assumption that the ancestors of *Welwitschia* had bisexual strobili. In *Ephedra* and *Gnetum* there are no such rudiments; but if these three genera had a common origin, the ancestors of *Gnetum* and *Ephedra* must also have been bisexual. None of them can have any pretensions, however, to the role of an angiosperm ancestor; angiosperm ancestors must have had truly bisexual strobili, whereas these plants have at best only a trace of former bisexuality; they are also excluded from consideration by their extremely reduced and specialised strobili and by the highly specialised structure of their conducting systems. We must therefore turn our attention to their possible

tions, in particular the acceptance of *Gnetum* and/or *Ephedra* as a precursor of the angiosperms, and involves so many strained interpretations of the facts that it is beyond the bounds of elementary logic and plausible reasoning. See, however, Heslop-Harrison (1958a), who discusses this problem from the point of view of developmental physiology of sex-expression in flowering plants.

ancestors; these, in all probability were some representatives of the order *Bennettitales*—see, for example, Arber and Parkin (1907), Zimmermann (1930, 1959), Takhtajan (1953, 1956).* This is the more likely, seeing that the *Bennettitales* are the only group of gymnosperms known to science in which we find bisexual strobili.

Since Saporta (in Saporta et Marion, 1885), the *Bennettitales* have often been proposed as possible ancestors of angiosperms, and in this connection the resemblance in structure between the strobili of the Mesozoic genus *Cycadeoidea* and the flower of *Magnolia* has often been pointed out. But this resemblance is wholly superficial; they are alike only in that both are bisexual and both consist of an elongated axis on which are arranged successively, and in the same order, protective bracts (perianth-members in *Magnolia*), microsporophylls and megasporophylls. But along with these few similarities there are profound differences. The microsporophylls (stamens) of *Magnolia* (as in other primitive angiosperms) are free and arranged spirally on the axis, but in the *Bennettitales* (and in *Welwitschia*) they are whorled and mostly connate. Moreover, according to Delevoryas (1965, 1968), the microsporangiate structure of cycadeoid cones seems to have been composed of a fleshy portion, continuous and cup-shaped below and with separate distal members, each with a fleshy appendage at the distal region. No less are the differences between their megasporophylls. Those of the *Bennettitales* are very reduced, simplified, stalk-like structures, sometimes very abbreviated, each bearing at its apex a solitary erect ovule. Between these stalk-like megasporophylls, and alternating with them, are sterile organs (interseminal scales) which appear to be modified sterilised 'megasporophylls'. These sterile scales, with their tightly packed expanded apices, form a kind of protective armour round the ovules. Protection of the ovules is achieved, therefore, in a very different way from that found in the angiosperms. The position of the ovules in *Welwitschia*, *Gnetum* and *Ephedra* is completely similar to that in the *Bennettitales*, but the number in each strobilus has been reduced to one.

* *Welwitschia* in particular has much in common with the *Bennettitales*. 'It is possible,' wrote Kozo-Poljanski in 1922, 'that *Welwitschia* is the last monstrous survivor of a once flourishing group of *Bennettitales*.'

It is evident that the extremely reduced and specialised stalk-like 'megasporophylls' of the *Bennettitales* and their descendants cannot be the starting-point of the angiosperm carpel. It is most unlikely that such 'megasporophylls' could have become transformed into the leaf-like megasporophylls of the primitive angiosperms, with their often numerous ovules in a more or less enclosed cavity. Once the protection of the ovules had been achieved by the interseminal scales, there would be no reason for the bennettitalean 'megasporophyll' to evolve into a closed carpel. Another special feature of the *Bennettitales* (and likewise of *Welwitschia* and its supposed relatives) is the presence in the ovule of a distinct micropylar tube, formed by the integument and serving for the reception of the microspores. In angiosperms there is no micropylar tube; the microspores are caught by the stigma of the megasporophyll, not by the ovule. Finally, the bennettitalean seed differs from that of primitive angiosperms in being exalbuminous, the embryo itself filling almost all the seed-cavity, and nutritive tissue being entirely absent or very scanty. All these facts show that the *Bennettitales*, and even more so *Ephedra*, *Gnetum* and *Welwitschia*, cannot have been the ancestors of the angiosperms. In fact the 'bennettitalean' hypothesis is now almost devoid of supporters, and the ranks of those supporting the 'gnetalean' hypothesis have noticeably thinned in recent years.

Though the *Bennettitales* cannot be the ancestors of the angiosperms, this does not mean there can be no phylogenetic links between them. It is quite possible that they are connected through a common ancestry, but from what common ancestor we cannot as yet say. If the secondary xylem of seed-ferns were more primitive and consisted of scalariform tracheids, their origin from this group would have seemed most likely; but not a single species with such xylem is known. In view of the extreme primitiveness of their leaves and reproductive structures, the absence of primitive wood seems strange. From what we know of the group, forms with primitive secondary xylem must have existed; among the palaeozoic seed-ferns there are even protostelic forms, such as *Tetrastichia* from the Early Carboniferous of Scotland, or *Heterangium*, a widespread Carboniferous genus. In the Early Carboniferous, and especially the Late Devonian, genera with scalariform tracheids in the secondary wood doubtless existed.* If they are discovered some day, one more gap in our knowledge of seed-plant evolution will have been filled.

If the *Bennettitales* and angiosperms did have a common origin from seed-ferns (Arber and Parkin, 1907), then their divergence must have begun very early, perhaps in the very earliest stages of strobilus evolution. Even earlier divergence—before the evolution of strobili began—cannot be excluded, though this seems less likely. More data, from new palaeontological discoveries, are needed for the final resolution of this problem.

We can thus conclude that angiosperms arose from some very ancient group of gymnosperms, which must have had primitive secondary xylem of scalariform tracheids at least in the early wood (see p. 32), and primitive bisexual strobili. The strobili must have been of a type that could have diverged to give rise to the primitive bennettitalean strobilus and the primitive angiosperm flower. On theoretical grounds they must have been rather large, terminal, and with an elongated axis bearing spirally arranged leavy bracts (perianth-members) and sporophylls. Both mega- and microsporophylls must have been leaf-like and pinnate. The microsporangia and ovules must have been numerous, the microsporangia free, and the ovules without a micropylar tube.

* Therefore many authors consider the origin of *Bennettitales* and cycads directly from seed-ferns quite possible; see, for example, Berry (1920), Chamberlain (1935) and Zimmermann (1959).

CHAPTER 4

ECOLOGY AND GENETICS IN FLOWERING PLANT EVOLUTION

The mysterious absence of fossil remains of the earliest flowering plants must, in our opinion, be a result of their ecological peculiarities. Most of the primitive angiosperms still surviving are, in fact, montane plants of tropical and subtropical areas. Therefore, it seems very likely that the first angiosperms (and their immediate gymnosperm precursors) were inhabitants of the mountains—see Berry (1934),* Vakhrameev (1947, 1952), Takhtajan (1948), Arnold (1947), Axelrod (1952, 1960), and Němejc (1956). If this conjecture is correct, then the striking gap in the fossil record is to some extent understandable. Mountain plants are usually found far from the areas of accumulation of sediments and buried plant remains; in other words, they grow in circumstances very unfavourable to fossilisation. But there is more to it than this; no less important is the fact that populations of small dimensions are characteristic of many mountain plants, and this in many cases leads to rapid rates of evolution (Takhtajan 1947, 1954a, 1957a; Alexrod 1952, 1960). A short digression into the realm of population genetics, the basis of the theoretical and experimental study of speciation, must now be made in this connection.

Concerning the development of population genetics we are in a large measure indebted to the eminent entomologist and geneticist S. Chetverikov, who published in 1926 his researches on some aspects of the evolutionary process from the standpoint of contem-

* Berry admitted the origin of flowering plants 'in the upland tropics' as one possible explanation, although not the only one. In a later work (Berry, 1945) he wrote, 'It may be that the real ancestors of the flowering plants may have dwelt remote from regions of fossilisation and may never have been preserved, although I regard such an explanation as a lame sort of an explanation.' The other authors cited later independently came to recognise the necessity of just such an explanation.

porary genetics. This classic work, devoted to an experimentally based analysis of free crossing and natural selection, was of great importance for the study of speciation and for evolutionary theory in general, as it paved the way for many further brilliant researches on population genetics in many countries. As a result of these researches it became clear that the population is the natural workshop in which the elementary processes that lead to evolutionary change take place.

By 'populations' we mean 'those discrete spatially isolated local groups or concentrations of organisms in which mating and interbreeding proceed more or less at random, or, in other words, in which something approaching panmixis in fact takes place; populations are the basic and elementary form of the community of existence of the individuals of each species' (Timofeef-Ressovsky, 1958). The population is thus characterised by free exchange of genes amongst its members. It therefore exists and acts as a 'unit of evolution'. However, the effective breeding population is in fact distinctly smaller than the size of the population as a whole. Interbreeding occurs predominantly between neighbouring individuals; those which stand at opposite ends of a sufficiently large population may be isolated from one another quite as effectively as if they were members of different populations (Stebbins, 1951). The size of a population may also be subject to seasonal and secular fluctuations, and in such circumstances the effective size of the population, according to Wright's calculations, corresponds to its minimal size. For these reasons, a large population may represent a system of 'effective populations' linked successively in space or time.

Evolution is a process subject to the laws of chance (a stochastic process), and therefore the regulating mechanism of natural selection functions effectively only when the populations consist of a sufficiently large number of individuals. This may be illustrated by a simple example. Let us suppose we toss a coin a number of times and wish to predict how many times it will show heads and how many times tails. A small number of throws will not suffice to elucidate the law operative for large numbers; but the larger the series of throws, the more constant the frequencies become, and after a sufficiently large number the frequency of heads (and correspondingly the frequency of tails) will approximate to one-half (i.e. 50%). This is because on a large enough scale the vagaries of chance are cancelled out, with the result that the operative laws become clearly apparent, whilst on the small scale they are obscured by individual fluctuations.

Natural selection as a regulating factor operates only on the large scale, for only in populations of a sufficiently large number of organisms does it effectively control the spread of mutant genes. In such cases it can even act too effectively; for, in spite of the great possibilities that exist for genetic variation in large populations, the rate of change of their genetic constitution is greatly retarded. In a large population a single mutation has usually no chance at all of spreading; and many repetitions of the same mutation are required to provide material for evolutionary change. 'If each species were represented by only a single, huge and largely panmictic population, then the differences that in fact arise in such populations would constantly tend to be eliminated through interbreeding, and this would greatly slow down the process of evolution' (Timofeef-Ressovsky, 1958).

In nature, however, species are extremely rarely represented by a single population, and far from all species consist only of large populations. In the majority of species, along with large populations there also exist more or less isolated small populations, and many species exist wholly as a number of small populations. These are especially characteristic of mountain (and particularly of high mountain) areas, of small lakes and small oceanic islands, and of the peripheral areas of species-distributions. The populations may be constantly small in number over a considerable period of time, or temporarily so, e.g. seasonally. Absolutely stable populations do not generally occur in nature; all fluctuate to a greater or lesser extent. The extent and frequency of numerical fluctuations may differ considerably in different species and under different conditions. Such quantitative variations are usually accompanied by an increase or decrease in the area occupied by the population. Chetverikov (1915) first pointed out the great evolutionary significance of quantitative fluctuations in populations, calling them 'waves of life'. Timofeef-Ressovsky (1958) proposed the term 'population waves', which more adequately conveys the significance of the concept. Another way in which small populations can arise is through migration. If during the migration of a species a new population is founded by merely a few individuals (or even by a single one) then obviously it will initially be small.

In small populations, especially those that consist of no more than a few hundred individuals, chance variations in the spread of mutant genes do not cancel one another out, and we therefore observe significant deviations from statistical expectation. There is the so-called 'genetic drift', or sampling effect, by which is meant a random spread of genes that is not, or is only weakly, controlled by selection, and is brought about by differential reproduction of the organisms. If a population (and consequently the number of interbreeding individuals) is sufficiently small, then their proportional increase cannot be effectively controlled by selection. In small, and especially very small, populations there are too few individuals for selection to act effectively within the population because too few events occur for the manifestation of the statistical law.

In whatever way a small population arises, whether through a sharp reduction in the numbers of a large population or by the separation of a small number of individuals from it in the formation of a new colony, its genotypic constitution does not as a rule reflect at all closely that of the original population. It is clear that, of the mutant genes present in low concentration in the heterozygous condition in the initial large population, only some will be represented in the random individuals that make up the small population. But the concentration of these genes in the new small population may be increased purely by chance, and thus the possibility of their transmission to the next generation may also be increased. If the population remains small, then this genetic drift will continue in future generations. Thus in small isolated populations purely chance fluctuations in the concentration of mutant genes predominate in deciding which shall persist into future generations. Some genes are accidentally lost and drop out of such populations, others, equally accidentally, increase in frequency and become fixed.

Therefore in small populations chance increases in the frequency of rare mutants are also possible. Such chance events or accidents of sampling have no real significance in large, freely interbreeding populations, but the smaller and the more isolated a population is, the more rapid will be the random loss or accumulation of mutants, and chance events therefore play a large role. As a result of such chance fluctuations the majority of mutant genes that arise individually in the population will disappear; some, on the other hand, will (unless very harmful) spread throughout the whole population until the population is stabilised and the process of drift concluded. If the influence of selection and mutation remains low, then as a result of inbreeding, a small isolated population will eventually, in a sufficient number of generations, reach the homozygous condition, a state of genetic homogeneity. Thus, if we have two allelles, *A* and *a* initially, one of them will inevitably be eliminated and the other will spread throughout the whole population. Which will be retained and which will be eliminated, however, depends purely upon chance; in different populations events may proceed differently, one allele being eliminated in one population, another in another, and so on. So, as a result of the cumulative effect of such chance variations, initially genetically similar populations may become genetically different, and thus assume different evolutionary potentialities. This in turn may lead to evolutionary divergence.

As mutational variation in very small populations is insignificant, under conditions of total isolation such populations will remain in a static condition, broken only from time to time by the chance fixation of rare mutants, and this will inevitably lead to degeneration and extinction. This is, of course, inimical to adaptive evolution. On the other hand, small isolated populations do have certain properties which, under favourable conditions, can promote rapid adaptive evolution. Such properties are the very rapid rate of change of the genetic constitution (up to the point of stabilisation), the possibility of the rapid fixation of rare favourable mutations, and the considerable genetic differences between isolated sister-populations which produce a high degree of inter-population polymorphism.

In spite of the low level of inheritable change in very small isolated populations, the spread of mutations proceeds therein much more easily than in large populations. While in large populations the spread of a gene may require tens of thousands of generations, in very small populations it may be very rapid and require but a few generations. This is because the elimination, or the fixation, of a mutant gene does require a certain minimal number of generations, but the fewer the individuals in the population, the smaller the required number of generations. This rapid action of drift can lead to positive results only in cases of fixation of rare favourable mutations; but in certain circumstances—to be considered more fully later—it may have important consequences.

Well-marked inter-population polymorphism is another important characteristic of small isolated populations. Genetically homogenous, such populations always differ from one another, although the differences are not as a rule adaptive. This non-adaptive character of the variation is due to chance fixation or propagation of mutations that are in most cases adaptively more or less neutral. Useful mutations, being very rare, have little chance of fixation in the absence of effective positive selection. Deleterious and neutral mutations are much more frequent; but while the deleterious ones are sooner or later eliminated by negative selection, the loss—or the fixation—of neutral or indifferent mutations is regulated solely by the chance process of genetic drift. Small neutral mutations occur apparently quite frequently and easily accumulate in the population. As a result, various forms of non-adaptive intraspecific polymorphism arise, which play an important part in the origin of interspecific differences, especially in plants. Non-adaptive polymorphism resulting from genetic drift in small isolated populations is especially characteristic of species growing in mountainous districts or on oceanic islands; while on limestone areas, under conditions favouring the isolation of small populations, a considerable number of narrowly localised endemic races and microspecies may arise.

The rapid spread of mutant genes in small isolated populations, and the genetic differentiation between such populations can, in certain circumstances, be of great evolutionary significance. Such circumstances are those that permit gene exchange to take place from time to time between the populations, as must often happen when the environmental conditions change. In other words, the chance differences which quickly arise between small local populations in temporary isolation, are of the greatest evolutionary importance. Occasional interbreeding between individuals of neighbouring populations not only appreciably increases their variability and hence their plasticity, but also results often in a well-marked heterosis (hybrid vigour). For these reasons, occasional hybridisation between representatives of isolated populations produces a great amount of material suitable for rapid adaptive evolution, which is brought about by selection both within and between the populations.

In all natural populations, large and small, both drift and selection act simultaneously, but whereas in a large population the effectiveness of selection is very great and the role of drift is extremely insignificant, in small populations selection is genetically very ineffective, and chance fluctuation in the concentrations of mutants predominates. Again, in large populations free interbreeding facilitates the realisation of many new combinations of genes but renders their fixation improbable, while in small populations the possibilities of recombination are limited by inbreeding but existing combinations may be rapidly fixed. In the former, evolution is strictly adaptive and slow; in the latter, it is usually non-adaptive and may proceed very swiftly, although this rapid rate is of short duration. The most favourable circumstances for rapid adaptive evolution are those in which selection and drift can operate in close conjunction and, as it were, supplement each other. This occurs when a large enough population (of the order of 10^5–10^6 interbreeding individuals) is broken up into a large number of local populations ('demes') temporarily isolated from one another, effectively interbreeding within themselves and from time to time exchanging genes through hybridisation. Useful combinations of mutations may arise through such exchange, and under the influence of selection can spread throughout the whole species or subspecies, while rare favourable mutations which may achieve chance high concentrations in the small populations can be spread to the other populations and become established by selection—see Wright (1931, 1948, 1949), Kolmogorov (1935), Dubinin (1940, 1966), Dobzhansky (1951), Simpson (1953), Waddington (1957), Dobzhansky and Pavlovsky (1957), Timofeef-Ressovsky (1958) and Grant (1963). Likewise, in the case of periodic fluctuation in the numbers of a population the advantages of large and small populations will be combined.

These conclusions of population genetics are of great importance, not only for their place in the general theory of evolution but also as an explanation of the systematic gaps in the palaeontological record. In palaeozoology they were first utilised by Simpson (1944), who came to the important conclusion that the absence of transitional forms between higher systematic units could not be explained merely by chance breaks in the geological record; and to aver that such forms never existed is even more unsatisfactory. As Simpson says in a later work (1953, p. 372), 'The lines leading to higher categories must often, or even usually, have had populations relatively quite small as compared with their descendants in the categories. This would markedly reduce the chances of their recovery as fossils.' According to Simpson, the initial process of separation of taxa of higher categories occupies far less time than the subsequent processes of adaptation, differentiation and dispersal. Vavilov (1926, 1951), Darlington (1963) and the zoologist Mayr (1954, 1963) paid special attention to the fact that in isolated peripheral populations evolution goes on at an accelerated rate. According to Mayr (1963, p. 513), new species and evolutionary novelties are usually produced by peripheral isolates, 'When a new species evolves, it is almost invariably from a peripheral isolate.' Mayr (1954) concludes that many puzzling features, especially of the palaeontological record, become explicable when the character of evolution in such populations is considered. Here we have a mechanism which enables us to explain the rapid origin of highly distinct new forms without conflicting with the facts of genetics. Palaeontologists have sometimes postulated some form of 'phylogenetic jump' to explain the absence of fossil intermediate links between certain higher systematic groups, but the genetic explanation of such 'jumps' has always been unsatisfactory. Now, of course, we know that there is no special method for the formation of higher systematic entities that differs in any way from ordinary 'microevolution'. According to present-day ideas, all evolutionary changes, from the smallest and most insignificant to the swiftest and most striking, occur on the population level. Evolution is nothing but the 'statistical transformation of populations' (Wright). The so-called 'macroevolution' is only the cumulative effect of rapid genetic changes on the population level. A higher entity is 'higher' because it *became* distinctive, varied (or both) to a higher degree and not directly because of characteristics it had when it was arising (Simpson, 1953).

I have attempted to explain the origin of angiosperms in terms of population genetics, and have thus concluded that the transitional groups between gymnosperms and angiosperms existed as small populations in which evolution went on at an accelerated rate (Takhtajan, 1947). Evidently they must have formed a system of small, semi-isolated local populations of mountain plants, many of them peripheral. Such systems of semi-isolated demes are common in mountain plants and provide the combination of drift and selection that is most effective in promoting rapid evolution. If this supposition is correct (and it is highly probable), then evolution in the immediate ancestors of angiosperms and in the earliest angiosperms themselves must have been very rapid, and must have facilitated the production of large adaptive modifications. Under such conditions, one of the ancient branches of gymnosperms might well have initiated a more progressive group of seed plants characterised by high adaptability and high plasticity. In my opinion, this group was unusually plastic because it arose by way of neoteny.

By *neoteny* is meant the genetically controlled persistence of the earlier stages of ontogeny and thus their evolutionary transformation into the adult stages of later generations. In other words, it is the cessation of ontogeny at an earlier stage than in the ancestral forms. As a result of such premature completion of ontogeny, the ancestral adult stage disappears from the individual life-histories of neotenic forms, i.e. ontogeny, as it were, cuts itself short. The neotenic transformation may affect the whole organism or only parts of it. 'Foetalisation' and 'juvenilisation' may be regarded as equivalents of neoteny in this wider sense; all these concepts embrace the idea of the evolutionary loss of the later stages of ontogeny and a general despecialisation of the whole organism or of parts of it.

The significance of neoteny in evolution lies in the simplification and despecialisation to which the neotenic forms are subjected. They are freed from the burden of specialisation borne by the ancestral adult stages. Such despecialisation, which is occasionally considerable, undoubtedly increases the possibility of evolution in new directions. Genetically this increase in plasticity depends upon the fact, demonstrated by Koltsov (1936), that abrupt neoteny involves a marked simplification of the phenotype alone; the genotype retains its complexity. This conservation of the genotype is of considerable significance for the further progressive evolution of the neotenic forms. This is because 'the reservoir of genes that are not now expressed in the neotenic forms (but which may mutate into genes that are expressed) gives them a high degree of variability and sometimes enables them to display in the future an exuberant outburst of progressive evolution'. Thus, in spite of phenotypic simplification, neotenic forms retain all the complexity of genotype amassed in the course of evolution, and this genetic potential facilitates development in new directions.

Neoteny may be understood genetically in terms of mutations of genes controlling the speed of ontogenetic development, and involves the reduction in activity of certain genes. In this respect it is important to note that the slowing-down activity of even a single gene (the 'suppressor-gene' of Koltsov) would affect the time of action of many genes and, as a consequence, the whole developmental process. Thus, according to Koltsov (1936, p. 520), 'It is possible that the impetus to the evolution of the *Diptera* was given by one original neotenic mutant gene which arrested development of the insect at a stage in embryogenesis at which the elaboration of the hind wings, mandibular mouth parts and antennae had only just begun; in the further evolution of the order this one suppressor-gene developed and fragmented into a number of different loci. Thus we have now, instead of one neotenic gene, a whole segment upon which the genes that delay the development of the different organs of the fly are situated. Back mutations which unlock the neotenic barriers can thus occur independently of one another at the different loci. In this way the results of experimental work on the genetics of *Drosophila* permit us perhaps to glimpse the nature of the original single mutant stimulus to neoteny which occurred millions of years ago and of which no clear palaeontological record has been preserved.'

Thus we may conclude that to explain neoteny there is no need to admit any radical genetic change of the Goldschmidt 'systemic mutations' type. From the evolutionary genetic point of view, *the significance of neoteny lies in the attainment of maximum phenotypic effect by means of minimal genotypic change.*

The evolutionary significance of neoteny has been fully realised by many zoologists; Garstang (1922), Koltsov (1936), Berrill (1955), Remane (1956), Hardy (1954), De Beer (1958), and many others have written of its importance. But while the significance of neoteny in the production of several main branches of the animal kingdom (including the mammals) has occupied the attention of many contemporary zoologists, very few botanists have acknowledged any major role for it in the production of the higher systematic entities of the plant kingdom. Neoteny is commonly invoked only to explain the origin of certain species or, more rarely, genera (*Phylloglossum*, *Welwitschia*) and families (*Lemnaceae*). As far as we know, no botanist has used the idea on such a scale as the zoologists have, although many botanists—Agnes Arber, for example—have accepted some role for neoteny in the morphological evolution of plants. In Arber's view (1937, p. 159), for example, 'retention of infantile characters has sometimes played a part in evolutionary history, and it is not inconceivable that something of the kind may also occur in the ontogeny of the individual', and analogous statements may also be found in the works of other authors—morphologists, anatomists and embryologists. I was thus led to attempt to apply the principle of neoteny to explain the evolution of certain large groups of higher plants. In a series of articles (beginning in 1943), I put forward the opinion that some major branches of the plant kingdom had developed by way of neoteny and subsequent progressive evolution. In particular, I came to the conclusion that neoteny gave the clue to the understanding of the origin of the *Bryophyta* from the *Rhyniophyta*, the angiosperms from the gymnosperms, and the monocots from the dicots.

The basic morphological peculiarities of angiosperms find their most likely explanation in the hypothesis of neotenic origin; both sporophyte and gametophyte bear the stamp of neoteny. Thus, as Arber (1937) concluded, 'the flower might indeed be described as corresponding to a vegetative shoot which remained in a condition of permanent infantilism' (p. 159). Likewise, Croizat (1947, p. 40) wrote, 'It is curious that it never seems to have occurred to "orthodox" morphologists that the *flower itself is an embryonal structure*, and that in most cases fertilisation reaches the flower in its embryonal stage.' In my view, the flower may be regarded as a neotenic form of the strobilus of the ancestral gymnosperms; as a whole and in its component parts it bears the marks of neoteny. For instance, the carpels of the most primitive angiosperms give the impression of being 'infantile' organs which correspond less to the adult sporophylls of the ancestral gymnosperms than to their juvenile stages. The evolutionary conversion of the open sporophyll into the carpel by folding and gradual closure along its middle vein could have occurred only in a juvenile stage. This process probably occurred the more easily because folded (conduplicate) vernation is characteristic of many angiosperms, the leaf in its young state being folded along its middle vein. It is very likely that the immediate ancestors of the angiosperms were also characterised by conduplicate vernation; in some of the most primitive angiosperms such as species of the section *Tasmannia* of the genus *Drimys* (Fig. 10, p. 61), and the genus *Degeneria*, the individual carpels are strongly reminiscent of young folded leaves (Bailey and Swamy, 1951). It is therefore very likely that the transformation of open megasporophylls into closed ones occurred at a stage in ontogenetic development at which they were still in a folded condition (Takhtajan 1948, p. 145). The stamens of angiosperms had an analogous origin, and undoubtedly the male and female gametophytes of angiosperms also had a neotenic origin, being completely devoid of all traces of gametangia. For details see Takhtajan, 1959 and 1964.

Traces of neotenic origin are also visible, though less obvious, in the vegetative parts of angiosperms. They are shown particularly in the morphology of the leaves of primitive angiosperms, which in some respects exhibit juvenile characteristics in comparison with the leaves of seed-ferns or even present-day cycads. The leaves of angiosperms arose, of course, from those of their ancestors, probably the seed-ferns. The leaves of the latter are undoubtedly the most primitive type known amongst the gymnosperms, and thus are most likely to have been the forerunners of the leaves of angiosperms. The development of the simple entire leaves of the primitive angiosperms could have occurred only as a result of marked simplification brought about by cessation of development at an early juvenile stage (Takhtajan, 1954b). Němejc (1956, p. 73), Axelrod (1960, p. 133) and Asama (1960) also concluded that the angiosperm leaf arose by way of neoteny. Němejc derived the angiosperm leaf from the archaic leaves of the primitive ferns. In his opinion, angiosperm leaves arose as a result of a progressive cessation of development at even earlier stages, this being associated with adaptation to an increasing dryness of climate. Axelrod also considered that the leaves of angiosperms were 'derived from naked phyllophores of a primitive fern or seed-fern alliance which were arrested early in growth, presumably as an adaptation to growing in more exposed upland regions.' Thus, in spite of differences on detail, all these authors in one way or another acknowledge the neotenic origin of the angiosperm leaf.

The exceedingly primitive structure of the wood of ancient angiosperms of the *Winteraceae* type is understandable in the light of the hypothesis of neotenic origin, their mature wood corresponding in its structure to the early wood of primitive gymnosperms, i.e. it likewise arose by neoteny (Takhtajan, 1961, p. 21).*

* The idea of the possible origin of the whole of the secondary wood of the *Winteraceae* type from the primitive early wood with scalariform tracheids of the hypothetical gymnospermous ancestors by way of neoteny was postulated by me in 1961, but was not developed or substantiated. Soon after, in his most interesting paper on paedomorphosis, or juvenilism (neoteny) in dicotyledonous xylem, Carlquist (1962, p. 44) showed that 'if paedomorphosis takes place in a wood, appearance of more primitive characters in the secondary xylem may result, because primary xylem tends to be a "refugium" for certain primitive characters, and transference of these characters to the secondary xylem would result in an

Thus the neotenic reorganisation evidently involved all the parts and organs of the angiosperm ancestors, and this was bound to have resulted in distinct breaks in the phyletic sequence of adult forms. The evolution of the angiosperms was therefore not only rapid (as a result of its occurrence in isolated peripheral populations), but also discontinuous as a result of neoteny.

In spite of their phenotypic simplification, the earliest angiosperms carried within themselves a large reserve of genetic potentiality that undoubtedly created a favourable genetic situation for rapid adaptive radiation. Thus, developing in the mountains and broken up into many small isolated populations, the earliest angiosperms found themselves under conditions most favourable to evolutionary radiation. And if we bear in mind that their evolution was closely tied to the evolution of insects and was based on the complex and peculiar mechanism of mutual selection, then the extraordinary speed of their initial differentiation becomes even more readily understandable.

admixture of primitive with specialised characters.' But the same may be true also in cases where late secondary wood with specialised tracheids is replaced, by way of neoteny, by more primitive early secondary wood with scalariform tracheids (see, however, Barghoorn, 1941).

CHAPTER 5

INSECTS AND THE FLOWERING PLANTS

Fundamentally the angiosperm flower differs little from a gymnosperm strobilus of the cycad or bennettitalean type, the main difference being that in the angiosperms (unlike the gymnosperms) the megasporophylls form the carpels, i.e. closed megasporophylls that contain the ovules. But if the structural difference is comparatively small, the biological difference is very great; inside the carpels, essentially new environmental conditions have been created for the growth of the ovules and for the process of fertilisation. As a result, many fundamental changes have taken place in the structure of the female gametophyte and in the process of fertilisation itself. We must now examine the circumstances that brought about the closure of the megasporophyll and the formation of the flower.

As Parkin (1923) noted, insects have played a prominent part not only in the evolution of the flower, as has been acknowledged by all since the work of Charles Darwin and Hermann Müller, but also in its very inception. After all, the most primitive types of flower, those of the *Magnoliaceae*, *Degeneriaceae* and *Winteraceae*, are characteristically entomophilous. And at the same time it may now be considered as firmly established that the flower was originally bisexual and that unisexual types have arisen as a result of reduction —see Chapter 2. It follows, therefore, that the ancestors of the angiosperms were also bisexual, for it is difficult to imagine the bisexual flower or strobilus arising from a unisexual one (we do not know of any examples of this kind), while the change from the bisexual type to the unisexual type is quite conceivable, examples being well known. Furthermore, cross-pollination in bisexual flowers is usually effected by animals (mostly insects) and only rarely by wind (as in grasses and sedges, but in these cases the flower usually has rudimentary petals, as represented by the lodicules in the grasses and by bristles in the sedges). From this one may conclude that the angiosperm flower arose most probably from a bisexual entomophilous strobilus, especially as among the earliest gymnosperms examples of bisexuality associated with entomophily are known within the *Bennettitales*.

Anthophilous insects must have played a decisive role in the transformation of a gymnosperm strobilus into the angiosperm flower. Insects visiting the strobili of the ancestors of the angiosperms found shelter therein, attracted by the abundance of pollen-grains on which they fed. At the same time, they facilitated cross-pollination. But they also ate the ovules, and thus damaged the flower. So naturally this primitive insect pollination must have led to protection of the ovules—see Arber and Parkin (1907), Gundersen and Hastings (1944), Gundersen (1950), Grant (1950a), Percival (1965) and Faegri and van der Pijl (1966). In various groups of seed plants this protection of the ovules has been brought about in different ways. In the *Bennettitales* the ovules were covered by special interseminal scales, which seem, however, not to have been an adequate defence against the insect pollinators of later epochs and in consequence probably led the plants to extinction in the course of the Early Cretaceous. A much more successful kind of adaptation for protecting the ovules was produced by the angiosperms. In the closed megasporophylls or carpels of the angiosperms the young succulent ovules were protected not only from the primitive unspecialised pollinators, but also from the suctorial insects of later epochs. Though they probably arose in the first place only as a defensive adaptation against the anthophilous insects, the carpels also gave excellent protection from many other harmful external influences, particularly that of atmospheric dryness. This does not, however, exhaust the biological significance of the new adaptation. As Golenkin (1927) pointed out in his interesting book *The victors in the struggle for existence*, the development of carpels made it possible for the angiosperms to simplify the structure of their ovules, and this in turn led to smaller ovules. The ovules found more reliable protection inside the carpel than in the primitive strobili of ancient gymnosperms, and thus no longer needed to elaborate such special defensive modifications as thickening of the integument, formation of a sclerotesta, and so on. As a result, the ovules became smaller and simpler, with a reduced gametophyte, and thus capable of much faster development. This was a very important advance, as it enabled the angiosperms to observe the greatest possible economy of material in the construction of the ovules and female gametophyte, and it also made possible the perfection of the process of pollination.

The development of carpels not only served to protect the ovules, thereby endowing them with the possibility of further evolution in new directions, but also led to the formation of the stigmatic surface, a further specialisation which was of great significance for the further evolution of pollination mechanisms. The acquisition of the stigma was undoubtedly a very great event in the evolutionary history of seed-plants. The function of catching the pollen-grains and stimulating their growth transferred from the micropyle of the ovule to the stigma of the carpel, and the loss of this function still further facilitated the simplification of the ovule. Moreover, the stigma can serve as a new barrier to 'undesirable' pollen and thereby assist cross-fertilisation.* Finally, the formation of carpels made possible both coenocarpy (union of the carpels) and angiocarpy (the enclosure of the carpels by other parts of the flower). This undoubtedly had a great adaptive significance. All

* According to Whitehouse (1950, 1960) incompatibility between pollen and carpel tissue of the same plant determined by a multiple-allele system may have been the primary cause of the evolution of the closed carpel and of the success of the angiosperms over their gymnospermous ancestors. But, as Baker (1963, p. 877) justly remarks, 'It is difficult to believe that the advantage of the outbreeding system of the angiosperms over systems employed by gymnosperms was great enough to be the whole explanation. After all, monoecism provides outbreeding while allowing complete interfertility between adjacent plants, and roughly two-thirds of the *Coniferales* are monoecious, with lesser proportions in other groups of gymnosperms.' He further points out that 'more recent studies of incompatibility systems have shown that several different kinds of multiple-allele systems exist, thus rendering the Whitehouse theory of a single origin of multiple-allele incompatibility unlikely'. Besides, as Crowe (1964) shows, 'incompatibility spans the entire plant kingdom'. She thinks, that 'the incipient angiosperms inherited incompatibility from their predecessors' and that they were hermaphrodites with a primitive ancestral inter-gametic incompatibility with many alleles at one controlling locus (p. 455).

these changes were also exceedingly important for the improvement of seed-dispersal mechanisms (see Corner, 1964).

Improved cross-pollination resulted, even in the case of the most primitive and archaic type of flower, and gave the ancient angiosperms a great advantage over the wind-pollinated gymnosperms and even over the *Bennettitales* with their still very primitive type of entomophily. The evolution of entomophily in angiosperms likewise began with very primitive forms. The first pollinators of angiosperms (and of their immediate gymnosperm ancestors) were not, of course, such specialised suctorial insects as butterflies and bees, but less specialised insects, as yet without a proboscis. Many authors have supposed them to be beetles—see Diels (1916), Kozo-Poljanski (1922), Grant (1950b), Porsch (1950), Leppik (1957, 1963), van der Pijl (1960, 1961), Eames (1961) and Baker (1963), amongst others—and thus that the first form of insect pollination was 'cantharophily'. Very many authors, from Delpino (1875) to van der Pijl (1960) and Eames (1961), have also pointed out that such ancient forms as many *Magnoliaceae*, *Winteraceae*, *Annonaceae*, *Nymphaeaceae* and *Eupomatia*, *Calycanthus*, *Illicium* and *Paeonia* are pollinated mainly by beetles. There is reason to think that cantharophily is shown also by so primitive an angiosperm as *Degeneria* (van der Pijl, 1960). The great antiquity of the beetles themselves is another argument in favour of the antiquity of cantharophily; in comparison with other groups of pollinating insects, the beetles are a very ancient order. They first appeared in the Early Permian (South Siberia and the Urals) and reached a high degree of development in the Triassic (Rodendorf and Ponomarenko, 1962). The anthophilous group *Nitidulidae*, on the other hand, is comparatively young—its development begins only with the Jurassic.

The *Diptera* and *Hymenoptera* are also rather ancient orders—their development begins with the Triassic—but they played a comparatively small role in the Mesozoic fauna, and the *Lepidoptera* are practically unknown from the Mesozoic. Hive-bees, bumble-bees, wasps, hover-flies (*Syrphidae*), bee-flies (*Bombyliidae*) and butterflies do not appear earlier than the Palaeogene (Danilevsky and Martynova, 1962; Rodendorf, 1962). Therefore, it is unlikely that they were among the most ancient pollinators of the angiosperms.

There are, however, other views. In Weismann's (1913) opinion, before the appearance of the butterflies and bees the then flat actinomorphic flowers with exposed nectar were frequented by a mixed company of caddis-flies, ichneumon-flies and sawflies. The development of the floral tube with concealment of the nectar later led to the exclusion of those visitors whose mouth parts were too short to reach the nectar and hence to the evolution of the proboscis, which elongated in parallelism with the floral tube. As a result of this co-ordinate evolution (co-evolution), there arose from the caddis-flies the butterflies, and from the ichneumon-flies the wasps and later the bees. However, the 'mixed company' must have been preceded by insects which did not require nectar as food (for nectar is absent in the most primitive angiosperms), but used the pollen. They were possibly beetles. The entomologist Malyshev (1964) has recently suggested that the first pollinators of angiosperms were some extinct type of wasp-like ancestors of the honey-bees (*Apoidea*); but the bees are a comparatively young group known only from Tertiary deposits, although they probably arose in the Cretaceous period. The first pollinators of the angiosperms must have existed long before the Cretaceous. Therefore, beetles remain as the most likely pollinators of the earliest angiosperms.

The initial unspecialised pollinators of the angiosperms—whether they were beetles or insects of some other kind—must have gradually faded into the background, as the leading role in the service of the angiosperms passed to highly specialised pollinators among the *Hymenoptera* (bees and wasps), *Lepidoptera* (butterflies and moths), *Diptera* (flies) and birds. This change in the pollinating animals probably occurred near the beginning of the Cretaceous period, or perhaps earlier, and had important consequences for the whole future of the angiosperms.

The development of specialised insect pollinators led to a very great advance in the process of cross-pollination, namely to an increase in the constancy of flower visitation (*Blutenstetigkeit*). Aristotle, in his *History of animals*, had already noted that a bee visited flowers of only one species on each flight, e.g. only flowers of the violet. Darwin (1876), who especially studied this question, affirmed that all species of bees, and some other insects, usually visit the flowers of one species for just as long as they can obtain food from it before transferring to another species. He wrote: 'That insects visit the flowers of one and the same species as long as they can is very important for the plant, as it facilitates cross-pollination of different individuals of one and the same species.' Such selective visiting of flowers of only one species of plant on each flight is met with in some beetles, but attains its highest development in *Lepidoptera*, *Hymenoptera* and *Diptera* (Kugler, 1955). The development of this selectivity required considerable specialisation of the insects' structure as well as development of the pattern of foraging behaviour. The more specialised pollinating insects are characterised by highly developed instincts of flower constancy and by a high capacity for creating new conditioned reflexes in response to stimuli in the environment, and in particular to flowers. At the same time, the development of constancy of visitation by insects demanded considerable specialisation of the flowers themselves, and in particular improved concealment of the nectar and pollen. And what was very important, it required standardisation of the dimensions of the flower and its component parts in line with the dimensions of the body and proboscis of the specific pollinator—see Berg (1956, 1958).

Flowers with many stamens and easily accessible food—pollen or nectar—are marked by the absence or weak development of constancy of visitation. Only species with such flowers are accessible to insects that do not enjoy the particular advantages of visiting one and the same kind of flower. In these cases the insects often fly from flowers of one species to those of another, and constancy, if it exists at all, is of an incipient nature. This is so, for example, for flowers such as those of the *Magnoliaceae*, *Nymphaeaceae*, actinomorphic *Ranunculaceae*, *Rosaceae*, *Liliaceae* and other polypetalous (i.e. dialypetalous) angiosperms. Things are quite different in the case of zygomorphic polypetalous, and especially gamopetalous, flowers. In such specialised flowers, the number of stamens is reduced and there is more opportunity for secretion and concealment of the nectar. For species with such flowers, definite constancy of visitation by the pollinators—usually insects—has become established, and in many cases this constancy holds good not only for each flight but for all flights as a whole. In such flowers pollination is carried out by bees, moths, butterflies and long-tongued flies, and in some cases by birds and bats—see Grant (1949, 1950c) and Kugler (1955). As was shown by Müller and Darwin, constancy of visitation is mutually beneficial to both plant and pollinator; the former is better pollinated with less expense of pollen, the latter can obtain food more easily. It represents the highest achievement of the co-ordinate evolution of angiosperms and their pollinators and is the result of a complex process of mutual selection.

The development of this constancy of visitation was of great importance in angiosperm evolution (see especially Grant 1949, 1963; Manning, 1957; Heslop-Harrison, 1958b). Besides facilitating cross-pollination between different individuals of the same species, it serves as an isolating mechanism that greatly reduces the possibility of interspecific hybridisation (Mather, 1947; Grant, 1963). It also reduces the possibility of interbreeding between *populations* that differ in the structure, colour or odour of their flowers (Grant, 1949). The change from primitive haphazard pollination to pollination exclusively by a particular insect (or other pollinator) subsequently led to the biological isolation not only of species and subspecies but of populations also. The isolation of a population is well-known to be a prelude to the formation of a new species. Therefore, the families of angiosperms with flowers more highly specialised for pollination show greater species differentiation of floral characters than those promiscuously pollinated by different insects (Grant, 1949, 1963; Stebbins, 1951). With the establishment of more specialised forms of pollination, the rate of angiosperm evolution greatly increased, and has continued gradually to increase (Grant 1949, 1963). Thus, the great number and diversity of the species and genera of angiosperms is largely a result of the isolating barriers occasioned by highly developed zoophily.

CHAPTER 6

THE FIRST FLOWERING PLANTS—A HYPOTHETICAL RECONSTRUCTION

Is it possible to depict, even in the most general terms, the appearance, structure and way of life of the angiosperms in the earliest stages of their evolution? This is the question which confronts everyone who is concerned with the problem of their evolution and origin, and although no fossil remains of the earliest angiosperms have been found, an answer can now in fact be given, albeit approximate and hypothetical. It may be obtained by the so-called hypothetico-deductive method, of which it therefore behoves us to say a few words. This deductive method differs in many respects from inductive methods of investigation. For inductive methods, the accumulation and classification of the greatest possible amount of factual data alone is important, but the deductive method requires a strict choice of data pertinent to the solution of the problem in hand, and the operation of a strict logical sequence in dealing with the facts.

Inductive methods do not enable us to rise above the level of classification of directly ascertainable information, but the deductive method permits us to put forward hypotheses to explain the connections between phenomena that are not open to direct observation. Therefore the hypothetico-deductive method is an important tool for the establishment of the histories of such diverse entities as the earth, the solar system, the plant and animal kingdoms, human society, languages and the universe itself. With its help, it is possible to reconstruct to some extent events of the past which we, at the present day, are unable to observe directly. Examples of the application of this deductive method are the various theories of the origin of the solar system and the attempts that have been made to determine past climates and the phylogenetic connections between organisms. The attempt to establish the characteristic features of the early angiosperms is also a typical example of the employment of the hypothetico-deductive method.*

Knowing the basic evolutionary pathways of the angiosperms and the main lines of specialisation of their organs and tissues, we may by extrapolation extend these lines mentally into the past to the lowest possible level of specialisation. This extrapolation is based on study of present-day primitive angiosperms, mostly of the order *Magnoliales*. Comparison with the gymnosperms, particularly the seed-ferns, cycads and *Bennettitales*, is also very important in many respects. Using such data, we can, by the hypothetico-deductive method, reconstruct step by step the appearance, morphology, anatomy, and in part even the ecology, of the early angiosperms. It must be remembered, however, that we can reconstruct in this way with any degree of certainty only the immediate ancestors of the primitive angiosperms that exist today. Nonetheless, even this reconstruction is of great interest.

This reconstruction of the ancestors of living primitive angiosperms (*Magnoliales*) depends on the truth of the assumption that they combined in one plant all the most archaic characters that are now found distributed among the different 'living fossils' in existence today. It is quite possible, of course, that the immediate common ancestors of present-day *Magnoliaceae*, *Degeneriaceae*, *Winteraceae*, etc. had features which are not present in any present-day angiosperms; we cannot be sure, for example, that they did not have the leaf-type of some seed-fern or other primitive gymnosperm. In other words, we cannot be completely certain that our reconstruction will take us far enough back into the past, but this, of course, is a limitation of any reconstruction of ancestral types based solely on analysis of living forms.

When all the most archaic features of contemporary primitive angiosperms (the *Magnoliales* and related orders), as established in numerous researches by various workers, are considered, we find the early angiosperms may be characterised as follows.

* For the role of hypothesis in science see especially K. R. Popper, *The logic of scientific discovery* (London, 1968).

Vegetative Organs

The first angiosperms were undoubtedly woody. That this is so and that the herbaceous habit was derivative, has been convincingly demonstrated by innumerable investigations and can scarcely be doubted. That the herbaceous stem is secondary in flowering plants was first proposed by Jeffrey in 1899, and this idea was developed in his book *Anatomy of woody plants* (1917); it was also independently proposed by Hallier (1901, 1905, 1912a). Later it was further developed and firmly substantiated by a whole series of morphologists and anatomists (see the literature in Takhtajan, 1964, p. 23). The reader will find the anatomical grounds for the derivation of herbaceous flowering plants from woody ancestors particularly in the textbooks of plant anatomy by Eames and MacDaniels (1947) and Esau (1965). The secondary nature of the herbaceous stem in flowering plants is also demonstrated by numerous data from comparative morphology and systematics. Whilst amongst the *Magnoliales* herbs are completely absent and amongst the *Laurales* extremely rare (only *Cassytha* and most species of *Chloranthus*), higher up the 'phylogenetic tree' one finds that they rapidly begin to predominate, and amongst the gamopetalous families they are clearly in the majority. Comparison of woody and herbaceous forms within different orders, families and genera leads to a similar conclusion; it is quite evident that as a rule herbs are more advanced in their structure than are related woody taxa. Finally, it is sufficient to recall that not one herbaceous gymosperm, living or fossil, is known, and thus the existence of herbaceous forms amongst the early angiosperms is improbable.

The early angiosperms did not, of course, resemble the stately and shapely trees of contemporary tropical rain forest. Rather, they were small trees with a weak crown of relatively few thick branches, like *Wielandiella* (*Bennettitales*) and *Cycas* (Hallier, 1901, 1912a, p. 148). In the opinion of Corner (1949a, 1964) the earliest stage of angiosperm evolution was the 'pachycaul' stage, with its characteristic unbranched or weakly branched massive sappy stems and soft wood. The earliest phase of pachycauly was, according to Corner, one of low, fleshy, unbranched, monocarpic trees with pinnately compound leaves. It is quite likely that the earliest angiosperms were characterised by some degree of pachycauly (like the *Bennettitales* and cycads), but amongst living primitive angiosperms the typical pachycaul form is not represented. Corner's examples of pachycauly (*Carica*, *Pandanus*, etc.) all belong to specialised groups.

The earliest angiosperms were evergreen plants (Hallier, 1901, 1912a), as are nearly all gymnosperms and most existing *Magnoliales*, and, as in the lower gymnosperms and the great majority of living primitive angiosperms, their leaves were alternate. Alternate phyllotaxis is known to be primitive, opposite and whorled leaf arrangements having been derived in the course of evolution (Nägeli, 1884, Zimmermann, 1959). The leaves of primitive living angiosperms are mostly simple, entire, pinnately nerved, coriaceous and glabrous. This indicates that the simple entire leaf with pinnate venation is primitive as far as present-day angiosperms are concerned (Parkin 1953; Takhtajan 1959; Eames, 1961) and it is very likely that the leaves of the earliest angiosperms were similar. But this is not certain—they may have been of a still more primitive type.

As in the present-day *Magnoliales*, the leaves of the earliest angiosperms were probably folded in bud, thus showing conduplicate vernation (Takhtajan, 1948).

Most phylogeneticists, Hallier (1912a) and Kozo-Poljanski (1922) among them, have considered that the leaves of the first angiosperms were stipulate. According to the latter: 'Stipules are especially well developed in angiosperms (like the tulip-tree), the great antiquity and primitiveness of which are shown in several ways—by the structure of the flower and shoot, by palaeontological data, and so on. In groups that are undoubtedly derived, they are either absent (e.g. *Asteraceae*) or modified (e.g. *Rubiaceae*). Typically stipules are small, with simplified vascular bundles, are without stomata, and are caducous or quickly wither. They thus give the impression of being rudimentary structures.' That the presence of stipules is correlated with the occurrence of tri- and multilacunar nodes and their absence with unilacunar nodes, has often been adduced as evidence in favour of the primitiveness of stipulate leaves (Sinnott and Bailey, 1914; Dormer, 1944, etc.). Unfortunately, the question cannot be so simply settled. Stipules are present only in the *Magnoliaceae* amongst the families of the *Magnoliales*; in all the other families (the majority of which have tri- or multilacunar nodes) there is no trace of stipules. Stipules are likewise absent in the majority of *Laurales* (except *Austrobaileycaeae* (Plate XII), *Chloranthaceae* (Fig. 18, p. 86) and *Lactoridaceae* (Fig. 19, p. 87)), and in the *Illiciales*. The absence of stipules in such undoubtedly very primitive families as the *Winteraceae*, *Degeneriaceae* and *Himantandraceae* makes it all the more difficult to regard the presence of stipules as a primitive feature. Although it has been suggested (by the present author among others) that the stipules of angiosperms may be the homologues of the 'aphlebia' of *Marattia* and the seed-ferns, this view has no real evidence to support it. The question of whether or not the earliest angiosperms had stipulate leaves must therefore remain open.

Stomatal Apparatus

The stomatal apparatus of flowering plants is characterised by a diversity of structure which has not yet been sufficiently studied, especially from the ontogenic point of view. It is therefore very difficult to establish which type is in fact the most primitive.

Stomata may be surrounded either by ordinary epidermal cells (the anomocytic type characteristic of *Nymphaeaceae*, *Ranunculaceae*, *Berberidaceae* and many other families), or by two or more subsidiary cells morphologically distinct from the other epidermal cells (paracytic, anisocytic, diacytic, actinocytic and other types)—see Metcalfe and Chalk (1950), Stebbins and Khush (1961), and Esau (1965). Developmentally, two types of subsidiary cell may be recognised—mesogenous and perigenous (in Florin's terminology). In the first type, both subsidiary cells and guard cells originate from the single common mother cell; in the second type, the subsidiary cells do not originate from the primary mother cell of the guard cells. (Cases are known, however, e.g. *Trochodendron* (Bondeson, 1952), in which one and the same stoma may have both mesogenous and perigenous subsidiary cells.) In recognition of this difference in development, Florin (1931) distinguished two basic types of stomatal complex, the halplocheilic (simple-lipped), in which the subsidiary (and, in general, the neighbouring) cells of the stoma are perigenous, and the syndetocheilic (compound-lipped), in which they are mesogenous. This terminology has now been applied to the flowering plants, although it appears to me altogether more expedient to replace it by the terminology proposed by Pant and Mehra (1964b), viz. 'mesogenous' in place of syndetocheilic and 'perigenous' in place of haplocheilic.*

According to Hallier (1908, 1912a) the primitive type of stomatal apparatus in flowering plants is devoid of subsidiary cells (the *anomocytic* type in present-day terminology). That the anomocytic type is primitive long seemed to me to be probably correct, but in recent years, working on my book *A system and phylogeny of the flowering plants* with the help of my assistant and pupil Margarita Baranova, I have surveyed stomatal slides of very many families and have come to the conclusion that the anomocytic type is derivative and that it arose from types with subsidiary cells. Only a few *Magnoliales* have the anomocytic type, while in more advanced families it is found much more frequently. It appears that in all *Winteraceae* (Plate XI/*3–4*), as well as in *Degeneria* and *Himantandra*, the stomata have subsidiary cells, while in the family *Magnoliaceae* the stomata of the great majority (including the most

* Pant and Mehra (1963, 1964b) consider that, as the term syndetocheilic was coined by Florin primarily to indicate the compound-lipped structure of the bennettitalean stoma, the development of which is unknown, and as it has been extensively used in this narrower sense, it is better for it to be used only for stomata of that particular type. The term 'haplocheilic' can then continue to be used topographically for mature stomata with the surrounding cells irregularly arranged. To distinguish different types of stomatal apparatus for which the ontogeny is *known*, Pant and Mehra suggest the use of the terms 'mesogenous' for stomatal apparati where the guard and subsidiary cells are derived by successive divisions from a single previous mother cell (meristemoid), and 'perigenous' for those in which neighbouring cells are derived not from a single previous guard-cell mother-cell but from surrounding protodermal cells which happen to lie around it. These authors correctly consider that the introduction of these new developmental terms will automatically clear up the confusion that has resulted from the use of the terms 'syndetocheilic' and 'haplocheilic' in a developmental as well as a topographical sense. At the same time, they will permit the terms 'syndetocheilic' and 'haplocheilic' to remain in use in literature on gymnosperm stomata where they have been extensively used mainly in the topographical sense.

primitive taxa) are characterised by the presence of subsidiary cells (Plate XI/*1–2*), whereas the anomocytic type is met with in, for example, so advanced a genus as *Liriodendron*. Which of the types of stoma with subsidiary cells—the mesogenous or the perigenous—is the more primitive?

In the evolution of the seed plants the perigenous (haplocheilic) type preceded the mesogenous (syndetocheilic) type (Florin, 1933, 1958), but the flowering plants most probably began with the mesogenous type. In support of this is the occurrence of the mesogenous type of stomatal apparatus in such primitive families as *Magnoliaceae* (Rao, 1939; Paliwal and Bhandari, 1962; Pant and Gupta, 1966) and *Winteraceae* (Bondeson, 1952). Moreover, the stomatal apparatus of the mesogenous *Magnoliales* is of the type known as paracytic (having one or more subsidiary cells on either side of the stoma, parallel with its long axis) (Plate XI). Thus the mesogenous paracytic type is the most likely basic type of stomatal apparatus in the evolution of flowering plants (Takhtajan, 1966).

In her work on the stomata and epidermis of *Magnolia*, Margarita Baranova (1962) notes a similarity of guard and epidermal cell structure between the primitive tropical species of *Magnolia* sections *Gwillimia*, *Lirianthe* and *Maingola* (and likewise *Manglietia*) and the *Cycadales*. She refers in particular to the presence of appreciable thickenings on the walls of the stomatal guard cells and epidermal cells, and she emphasises especially the similarity of the epidermis of several species of the sections *Gwillimia* (*M. championii*, *M. paenetalauma*) to that of the genus *Stangeria* (*Cycadaceae*), for which the same type of sinuosity of outline of the epidermal cell walls is characteristic. She also notes the presence in *Manglietia forrestii* (Plate XI/*1*) of characteristic cuticular striations radiating outwards from the bases of the hairs of the lower epidermis and reminiscent of such striations in *Stangeria*. It is certainly possible that this similarity to a member of the *Cycadales* is purely superficial and is a result of convergence.* But, apart from this, I consider that

* In this connection, it is interesting to find that Pant and Gupta (1966) note the similarity of the thickened lamellae on either side of the pore of the stomata of the species of *Magnolia* and *Michelia* that they studied to the cutin or lignin lamellae in typical gymnospermous stomata.

Baranova is correct in regarding the presence of thickenings on the walls of the guard, subsidiary and epidermal cells of primitive *Magnoliae* as a primitive feature; this is made more probable by the fact that her researches have shown that such thickenings are absent in all *Magnoliae* from temperate climates, including all the polyploids. The primitive structural features indicated by Baranova are found only amongst the diploid species.

Further, more extensive and detailed studies in comparative and developmental stomatography will show to what extent these preliminary tentative conclusions are correct.

Nodal Anatomy

There are three basic types of node in angiosperms, the trilacunar, the multilacunar and the unilacunar. It is natural to ask which of these three types is the most primitive and which type may have occurred in the earliest angiosperms. To answer this question, we must first consider the gymnosperms and ferns. All ferns, and the vast majority of gymnosperms (including the seed-ferns and *Bennettitales*), have unilacunar nodes, but in the *Cycadales* and *Gnetum* they are multilacunar (Pant and Mehra, 1964). Of these two types, in gymnosperms the unilacunar is undoubtedly more primitive, and it is generally agreed that the multilacunar type arose from it in course of evolution. But, however clear to us the picture of nodal evolution in the gymnosperms may be, it gives us very little indication of the primitive type in angiosperms. The angiosperms may have arisen from either unilacunar or multilacunar ancestors, and the plants intermediate between gymnosperms and angiosperms may have had either one type or the other. Furthermore, in angiosperms there is a third type, the trilacunar, unknown in gymnosperms. Therefore, the question as to which of these three types is primitive must be decided by the comparative anatomy and phylogeny of the angiosperms themselves.

Sinnott (1914), in an interesting publication on nodal anatomy and its significance in the classification of angiosperms, concluded that the trilacunar node was primitive and the unilacunar and multilacunar types derivative, and this idea quickly gained general acceptance. Not until the 1940s and '50s (especially the latter decade) did it begin to be revised.

Ozenda (1949) concluded, on the basis of studies of the nodal anatomy of the *Magnoliales*, that the primitive type was the multilacunar, not the trilacunar. In his view, the three types known in angiosperms form a regressive series, from the multilacunar through the trilacunar to the unilacunar. Later, in the 1950s, Marsden and Bailey (1955), Canright (1955) and Bailey (1956) considered the primitive node to be the unilacunar type with two discrete leaf traces, the so-called 'fourth type'. This new concept of nodal evolution was supported by many other anatomists—Foster and Gifford (1959), Esau (1960, 1965), Eames (1961), Carlquist (1961), and others. It was based on the fact that the unilacunar node with two distinct traces is characteristic not only of some gymnosperms and ferns (as was well-known earlier*), but also occurs in certain dicots (*Laurales*, including *Austrobaileya*, *Lactoris*, *Amborella* and *Monimiaceae*, certain *Verbenaceae*, *Lamiaceae* and *Solanaceae*). Also, it is repeatedly found in the cotyledonary node of various angiosperms. Marsden and Bailey considered all these facts to indicate that the primitive nodal type of the ancestral angiosperms was more likely to have been the unilacunar type with two discrete leaf traces than the trilacunar type. The unilacunar node with a single trace arose, in their opinion, by fusion of the two leaf traces, and the trilacunar and multilacunar types successively by the addition of lateral traces arising from new gaps in the vascular cylinder. In his article 'Nodal anatomy in retrospect', Bailey (1956) concluded that we could no longer think of the unilacunar node of dicotyledons as having arisen by reduction from the trilacunar; in his opinion, 'during early stages of the evolution and diversification of the dicotyledons, or of their ancestors, certain of the plants developed trilacunar nodes, whereas others retained the primitive unilacunar structure'. Canright (1955), Eames (1961) and several other anatomists have even more strongly favoured the primitiveness of the unilacunar node with two traces,

* Though not for many of them. As Pant and Mehra (1964a) have recently shown, a vast majority of the ferns and gymnosperms do not show two-trace nodes, as Marsden and Bailey (1955) suggested.

which they consider the basic type in the evolution of angiosperm nodal structure. But there are also objections. Thus Benzing (1967) has recently pointed out that the occurrence of plants with two-trace unilacunar nodal structure proposed as primitive by Marsden and Bailey is limited to a few families characterised by derived decussate phyllotaxy and many specialised floral characters. He comes to the conclusion that either the unilacunar node with one trace or the trilacunar node with three traces is more likely to be primitive in the *Angiospermae* than the fourth type (p. 820). In his opinion, the multilacunar type, which is restricted to only three primitive families (*Magnoliaceae*, *Degeneriaceae* and *Eupomatiaceae*), 'probably represents a specialised type' (p. 819).

Marsden, Bailey, Eames and others are apparently correct in asserting that the uneven numbers of leaf traces arose from an even number. On the other hand, the number of *gaps* is always uneven (1, 3, 5, 7, 9, etc.). Thus it follows that the basic type could only have been one with an *uneven* number of *gaps* and an *even* number of leaf traces. It could have been either a unilacunar node with two leaf traces (the fourth type), or a tri- or multilacunar type from the central gap of which arose two traces (this we may call the 'fifth type'). Consideration of the nodal anatomy of the most primitive groups of angiosperms convinces me that it was more likely the hypothetical fifth type than the fourth type. Thus all families of the *Magnoliales* in the strict sense have tri- or multilacunar nodes. In *Magnoliaceae* and *Eupomatia* they are multilacunar, in *Degeneria* pentalacunar, and in *Himantandra*, *Winteraceae*, *Annonaceae*, *Canellaceae* and *Myristicaceae* they are always trilacunar (Ozenda, 1949; Bailey and Smith, 1942; Bailey and Nast, 1944; Canright, 1955). It is also very significant that in the *Magnoliaceae* and in *Degeneria*, as shown by Swamy (1949), the central cotyledonary gap has two traces.

In their nodal anatomy the *Trochodendrales* have much in common with the *Magnoliales*; in *Trochodendron* the nodes have five to seven lacunae, and in *Tetracentron* they are trilacunar (Bailey and Nast, 1945b). Nodes with three, five, seven or more gaps are also characteristic of such relatively primitive orders as *Nymphaeales*, *Piperales*, *Ranunculales*, *Rosales* and *Dilleniales*, though they are found also in some more advanced orders (including the *Asterales*). In some orders, such as the *Celastrales*, it is possible to follow the transition from the trilacunar to the unilacunar type, which occurs along with general specialisation of the vegetative organs. It is particularly well shown in the family *Icacinaceae*—see Bailey and Howard (1941). One may see the same thing in the series *Dilleniales –Theales* and in the order *Santalales*. It is also significant that gamopetalous dicots (with the exception of the *Asteraceae* and the *Plantaginaceae*) are characterised by unilacunar nodes.

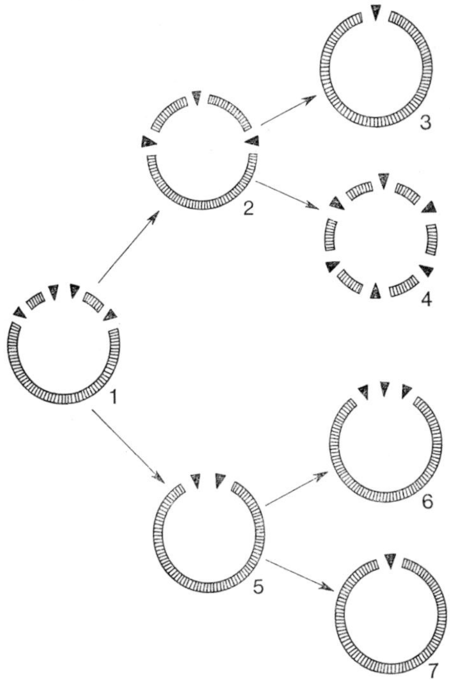

Fig. 3. Probable course of evolution of nodal structure in dicotyledons from a primitive tri-multilacunar type: *1*, tri-multilacunar node with a double trace at the median leaf gap; *2*, trilacunar node with three traces; *3*, unilacunar node with one trace; *4*, multilacunar node with many traces; *5*, unilacunar node with a double trace; *6*, unilacunar node with three traces; *7*, unilacunar node with one trace. (After Takhtajan, 1964.)

The double-leaf trace, moreover, is not confined to plants with unilacunar nodes. A double structure is always more or less evident not only in the traces of unilacunar nodes, but also in the traces from the central gap of tri- and multilacunar nodes; this is especially true of the cotyledons (where it has long been known), but holds also in many cases for the central bundle of the leaves of the adult plant. *Degeneria* is particularly interesting in this respect; at the trilacunar cotyledonary node two discrete leaf traces are clearly visible arising from the central gap (Swamy, 1949).

To me it seems that all these facts lead to one conclusion—that the basic type of angiosperm node was one with three or more gaps. To indicate the basic type more precisely is difficult at present; but if we take into consideration the complete absence of the trilacunar node in gymnosperms, and the presence of multilacunar nodes in *Cycadaceae* and *Gnetum*, then we must conclude that the nodal structure of the earliest angiosperms was more likely multilacunar than trilacunar. Possibly it was of the *Degeneria* or *Magnolia* type but most likely with two discrete leaf traces arising from the central gap. This hypothetical synthetic fifth type then gave rise to all the types of nodal organisation known to us (Fig. 3).

Anatomy of the Stem Conducting System

The presence amongst living angiosperms of genera and families with primitive vesselless wood indicates that the earliest angiosperms must have had a very primitively organised water-conducting system in their stems. The tracheids had scalariform bordered pits as in the living genera *Trochodendron* and *Tetracentron* (Fig. 4).

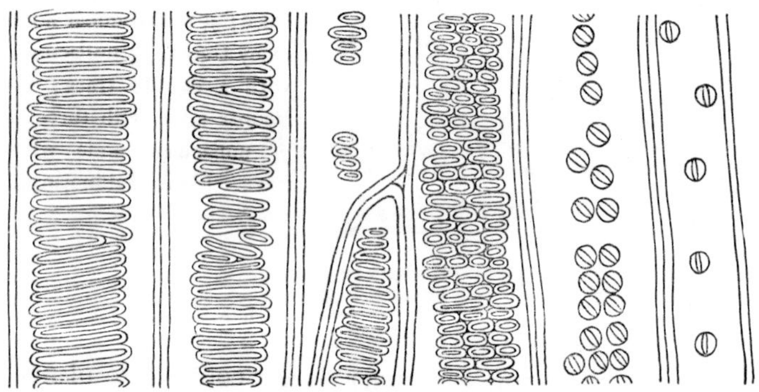

Fig. 4. Radial section of the primitive vesselless wood of *Tetracentron sinense*, × 100. (After Takhtajan, 1959.)

The phloem was also undoubtedly very primitive. As comparative anatomical studies of the phloem from Hemenway (1913) onwards have shown, the sieve elements of primitive angiosperms are long and narrow with very oblique end walls, as, for example, in *Drimys*. This is in agreement with the finding that the sieve elements in ferns and gymnosperms are long and pointed with no pronounced differences between the side and end walls. The absence of companion cells in the phloem of gymnosperms and ferns gives us good reason to suspect that the earliest angiosperms were also devoid of them.

From study of the comparative anatomy of gymnosperms and angiosperms it is also possible to draw certain conclusions as to the ancestral type of storage tissue—wood parenchyma and rays—in the xylem (Fig. 5). Wood parenchyma (occurring as longitudinal

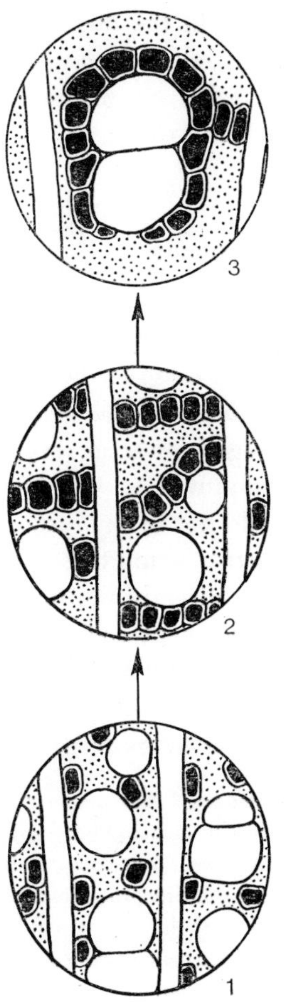

Fig. 5. Evolutionary relationships of the main types of wood parenchyma: *1*, diffuse parenchyma; *2*, apotracheal banded parenchyma; *3*, vasicentric parenchyma. (After Takhtajan, 1964.)

parenchyma strands) in early angiosperms was either very scanty (Hallier, 1908, 1912a) and apotracheal (independent of the tracheal elements in its distribution) or, more probably, was absent, as in seed-ferns, *Bennettitales* and cycads. Carlquist (1962) considers absence of parenchyma as primitive. Rays (radial tracts of storage parenchyma) were undoubtedly present, as they are in gymnosperms. They must have been of the heterogenous type, consisting of a combination of multiseriate rays with high uniseriate 'wings' and uniseriate rays composed of high upright cells. Multiseriate rays must have been of the heterocellular type, consisting of two different types of cell, upright cells (elongated in the direction of the stem) and procumbent (radially elongated) cells. Such rays are met with in many living angiosperms with relatively primitive wood (Fig. 6) (Kribs, 1935; Metcalfe and Chalk, 1950; Eames, 1961; Esau, 1965.)

General Floral Structure

Study of the flowers of primitive living families can give us a good idea of the general floral organisation of the earliest angiosperms. A. Braun (1875) towards the end of his life came to the conclusion that the most primitive flowers were those of the *Magnoliaceae*, *Alismaceae* and similar families. Later, Nägeli (1884, p. 519) wrote, 'The *Ranunculaceae*, which stand at the summit of A. P. de Candolle's system as the most perfected of plants, do not in any respect reach so high a phylogenetic level. In their construction these plants are very primitive, as the flowers are borne directly terminal on the leafy branches. Likewise, in their arrangement and construction the androecium and gynoecium are primitively organised, as the stamens and carpels are usually numerous, spirally arranged, and free.'* Later, these ideas were developed in more detail in the

* 'Die Ranunculaceen, die als die vollkommensten Pflanzen an der Spitze des Systems von A. P. de Condolle stehen, erreichen in keinem einzigen Merkmale einen hohen phylogenetischen Rang. Der Aufbau des Pflanzenstockes bleibt auf der untersten Stufe, indem die Blüthen an den Laubblattsprossen terminal sind. Ebenso stellen das Androeceum und Gynaeceum bezüglich Stellung und Verwachsung niedrige Bildungen dar, da die Staubgefasse und Carpelle gewöhnlich zahlreich, schrauvenständig und nicht mit einander verwachsen sind.'

works of Hallier, Bessey, Arber and Parkin, and their numerous followers. We have, in consequence, every reason to believe that the earliest angiosperms were characterised by large solitary

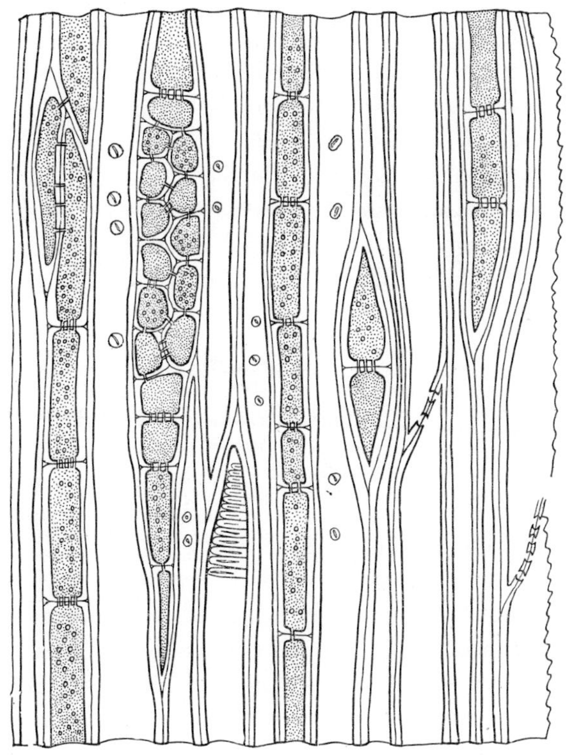

Fig. 6. Primitive rays of *Illicium parviflorum*. (After Takhtajan, 1948.)

flowers terminal on the leafy branches—see Hallier (1912a) and Parkin (1914).

The flowers of the earliest angiosperms must have had an elongated axis (receptacle) with an indefinite and variable large number of free parts arranged spirally upon it. They were radially symmetrical (actinomorphic) and bisexual. Such flowers were still very similar to the strobili of the cycadicaean gymnosperms.

Perianth

In the most ancient angiosperms there was very probably no corolla (Hallier, 1912a, p. 199), the perianth consisting entirely of modified bracts, as in the *Bennettitales*. The perianth was apparently composed of many thick leaf-like sepals arranged spirally (as in *Paeonia*) and (as in *Calycanthaceae*) gradually passing below through a series of intermediate structures into the foliage leaves. The sepals probably had palmate venation and three leaf-traces, as in the *Ranunculaceae*. The first petaloid organs were probably 'bracteopetals', that is to say they originated from bracts. In the *Magnoliales*, as well as in *Illiciales*, *Nymphaea* and *Victoria*, *Nandina* and *Paeonia* all the perianth parts (even though the perianth is differentiated into calyx and corolla) are morphologically equivalent and of bract origin (Ozenda, 1949; Hiepko, 1965).*

Pollination

The flowers of the earliest angiosperms were most probably insect-pollinated. Henslow (1888), in his now almost forgotten book *The origin of floral structures through insect and other agencies*, wrote, 'There is little doubt but that all wind-fertilised angiosperms are degradations from insect-fertilised flowers' (p. 266). Later, the idea of entomophily as primitive and anemophily as derivative in angiosperms, first convincingly advanced by Robertson (1904), gained general acceptance, and has now been confirmed by numerous investigations of the morphology and biology of the flower—see van der Pijl (1960) and Eames (1961). Robertson considered that insects probably visited the flowers of primitive angiosperms mainly because of their nectar, but the majority of authors consider that they were first attracted by pollen, not nectar. The latter suggestion more probably explains the origin of entomophily; it seems extremely unlikely that secretion of nectar preceded the visiting of flowers by insects.

Stamens

Comparative study of the stamens of angiosperms leads to the con-

* Hiepko (1965) makes an important contribution to our understanding of the origin of corolla in the primitive angiosperms.

clusion that the primitive type of stamen was a broad laminar structure (Fig. 7), undifferentiated into filament and connective, and produced beyond the microsporangia (Hallier, 1901, 1912a; Arber and Parkin, 1907; Kozo-Poljanski, 1922; Parkin, 1923, 1951; Bailey and Smith, 1942; Ozenda, 1949, 1952; Canright, 1952;

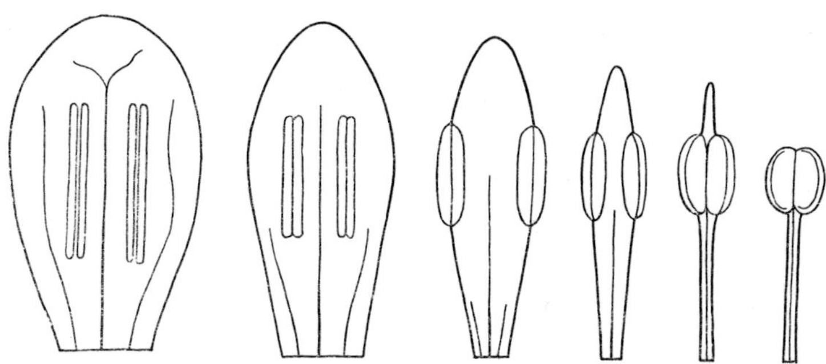

Fig. 7. Main trend of evolution in stamens from the most primitive broad laminar type. (After Takhtajan, 1954.)

Moseley, 1958; Eames, 1961). The stamens of *Degeneria* (Fig. 8), *Himantandra* and some *Magnoliaceae* are especially primitive. The majority of primitive laminar stamens, like those of *Degeneria*, are characterised by three main veins and three traces. The stamens of the earliest angiosperms were also in all probability broad leaf-like microsporophylls with a three-trace, three-gap system.

The stamens of most angiosperms have typically four microsporangia (the basic number). In primitive groups, the microsporangia are long and narrow, but have shortened in the course of evolution. In all primitive families with laminar stamens the microsporangia are superficial, i.e. sessile on the surface of the microsporophyll, *not* on the margins. It is interesting to note, however, that while in the *Magnoliaceae* (except *Liriodendron*), *Austrobaileyaceae* and *Nymphaeaceae* the microsporangia are situated on the upper (adaxial) surface (and therefore the stamens are introrse), in the *Degeneriaceae*, *Himantandraceae*, *Lactoridaceae*, *Annonaceae*, *Calycanthaceae*, *Lardizabalaceae*, *Belliolum* (*Winteraceae*) and *Liriodendron* they occupy the lower (abaxial) surface (the stamens being extrorse). The presence of these two diametrically opposed types in such closely related families as, for example, *Magnoliaceae* and *Degeneriaceae* is difficult to explain. If we allow these families a common origin (and it is difficult not to), then theoretically there are only three possible situations: (i) that the adaxial type is derived from the abaxial; (ii) that the abaxial is derived from the adaxial and (iii) that both are derived from a third type, which must have been the marginal. Since the abaxial position is evidently

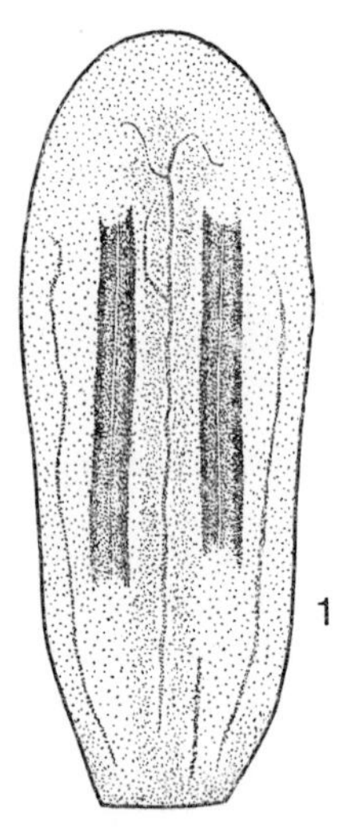

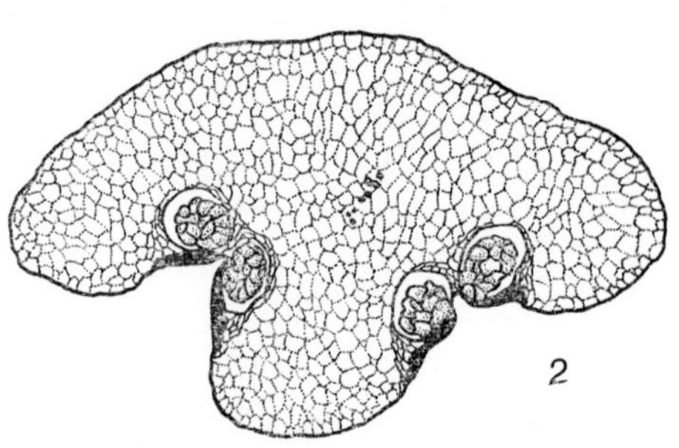

Fig. 8. The primitive stamen of *Degeneria vitiensis*: *1*, stamen re-expanded and cleared, showing sporangia and median and lateral veins, × 14; *2*, transverse section of stamen showing four deeply embedded sporangia and four short arcs of endothecium, × 60. (After Bailey and Smith, 1942.)

characteristic of the majority of ferns and the cycads, Hallier (1901, 1912a) considered it more primitive than the adaxial. But marginal and terminal sporangia are also found amongst the ferns, and marginal (many seed-ferns) and adaxial (*Williamsoniaceae*) amongst the gymnosperms. Furthermore, as Eames (1961) pointed out, many instances are known of the shift of microsporangia from the adaxial to the lateral and abaxial positions. In some cases, e.g. the palms and *Polygonaceae*, the shift from the adaxial to the abaxial position occurs during ontogeny (the stamens are introrse in the bud but extrorse in the flower); in others, as in the *Magnoliaceae* and *Cabombaceae*, the shift appears to have occurred in the process of evolution. In Eames's opinion, 'it is apparent that, at least in many families, the primitive position for microsporangia is adaxial', from which, however, he did not conclude that the abaxial position was derived from the adaxial. On the contrary, he considered that it seemed 'unlikely that the abaxial position in the highly primitive stamens of *Degeneria*, *Himantandra* and *Lactoris* is secondary—that these stamens are highly specialised in sporangium position. It is possible that both positions are primitive, a retention in the stamen of a morphological structure of an ancestral taxon' (p. 123). While agreeing with this conclusion of Eames, I must, nevertheless note that, if we accept the common origin of *Magnoliaceae*, *Degeneriaceae*, *Himantandraceae* and similar families, we must come to the logically inescapable conclusion that both adaxial and abaxial types have been derived from a common ancestral type, which could only have been the marginal. Thus we conclude that in the (perhaps very distant) ancestors of the *Magnoliales* the microsporangia were marginally situated on the microsporophylls. Nevertheless, it is still difficult to see clearly how the laminal position of the angiosperm microsporangia could have arisen from the marginal; more research and information are needed on this question.

Many authors, among them Ozenda (1952), Canright (1952), Moseley (1958) and Eames (1961), consider that the immersion of the microsporangia in the tissue of the microsporophyll is a primitive feature. In *Degeneria* and *Himantandra* the microsporangia are deeply sunk in the tissue of the stamen, as they are in the *Magnoliaceae* (except *Liriodendron*) and *Victoria amazonica*, and it is possible that the earliest angiosperms were also characterised by immersed sporangia, the shift from the marginal to the laminal position having already occurred.

Pollen Grains

It is very important to establish the basic type of the wall (sporoderm) of the angiosperm pollen grain. According to Hallier (1912a), the primitive type was characterised by one 'germinal pore', by which he apparently meant 'aperture' and not a pore in the strict sense of the word. Later it was shown that the most primitive angiosperm pollen grain was a type with one polar germinal furrow (distal colpus or 'sulcus') in the spore wall—see Wodehouse (1953), Bailey and Nast (1943a), Takhtajan (1948, 1959) and Eames (1961).

Such monocolpate pollen grains still have a continuous aperture membrane devoid of special openings ('ora') in the exine for the emergence of the pollen tube. Monocolpate pollen is characteristic of the *Magnoliales* (Fig. 9) and some other primitive dicots, of the

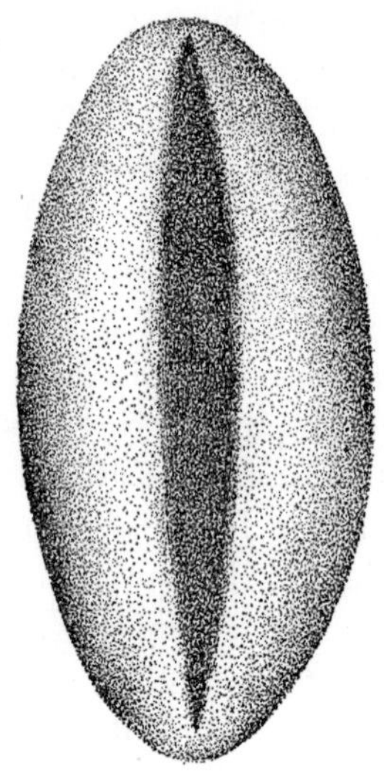

Fig. 9. Primitive monocolpate pollen grain of *Magnolia grandiflora*, greatly enlarged. (After Takhtajan, 1948.)

majority of monocots, and of many gymnosperms (including the *Bennettitales*, cycads, *Welwitschia* and *Ginkgo*). The pollen of the earliest angiosperms was in all probability of the magnoliaceous or cycadaceous type.

Carpels

The primitive carpels were large and leaf-like, containing a fair number of ovules (Prantl, 1887, Hallier, 1901, 1912a; Arber and Parkin, 1907; Eames, 1931, 1961). The number has in most cases gradually decreased in the course of evolution, although in many families, such as the *Ericaceae*, *Orobanchaceae*, *Orchidaceae*, etc., it has on the contrary greatly increased. Primitive carpels are reminiscent of young leaves folded adaxially along the midrib (Figs. 10, 11); such conduplicate carpels are characteristic of certain primitive living angiosperms, e.g. the *Drimys* sect. *Tasmannia*

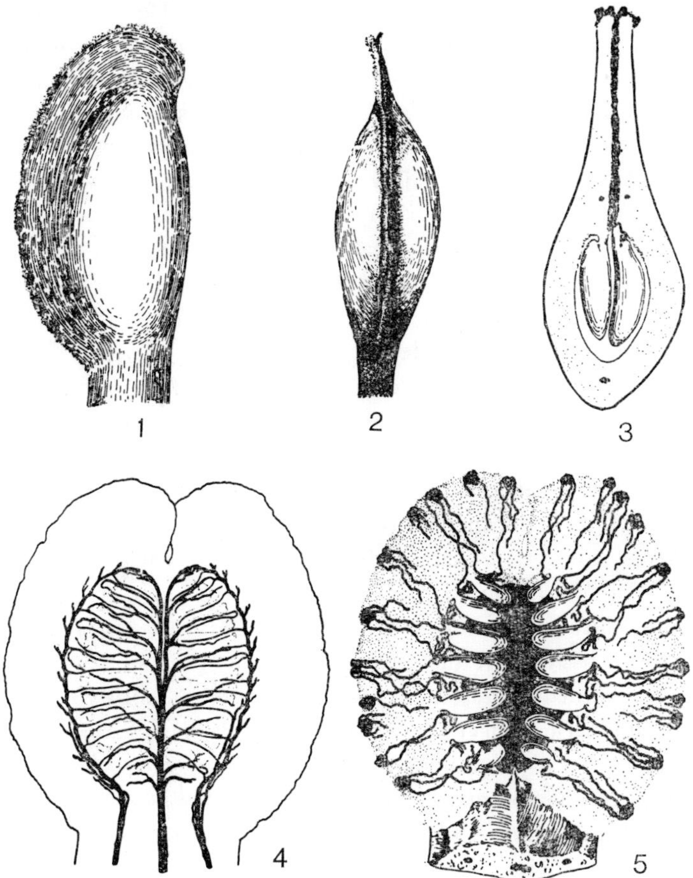

Fig. 10. The primitive conduplicate carpel of the *Drimys* sect. *Tasmannia* type: *1*, side view; *2*, ventral view, showing the paired stigmatic crests; *3*, transverse section, showing penetration of pollen tube; *4*, cleared, unfolded lamina, showing vasculature; *5*, unfolded lamina, showing primitive lateral-laminar placentation, distribution of glandular hairs (stippled), and the course of pollen tubes. (After Bailey and Swamy, 1951.)

(Fig. 10) and *Degeneria* (Bailey and Swamy, 1951) and to a lesser degree the family *Magnoliaceae* (Canright, 1960) and some others. The presence of a petiole-like stipe is a feature of the carpels of many primitive angiosperms; it is especially well shown in *Degeneria* (Fig. 15/4, p. 77), *Drimys*, *Bubbia*, *Austrobaileya*, *Calycanthus*, certain

Annonaceae, many *Ranunculaceae*, *Cercidiphyllum* (Fig. 25, p. 101), *Euptelea*, *Paeonia*, many *Dilleniaceae* and even some monocots (Bailey and Swamy, 1951; Eames, 1961). In the early angiosperms the carpels were not completely closed at the time of pollination, as is still the case even in living forms like *Drimys* sect. *Tasmannia* (Fig. 10) and *Degeneria* (Bailey and Smith 1942; Bailey and Nast, 1943b; Bailey and Swamy, 1951), and also *Coptis* (Eames, 1961); *Platanus* (Boothroyd, 1930), certain primitive monocots (Eames, 1961) and others. They were still as a consequence 'hemiangiospermous' plants. Primitive carpels are also characterised by ternate venation, as for example, those of *Degeneria*, *Himantandra*, sect. *Tasmannia* of *Drimys* (Fig. 10), *Magnoliaceae*, many *Ranunculaceae*, etc. (Eames, 1931; Bailey and Smith, 1942; Bailey, Nast and Smith, 1943; Bailey and Nast, 1943b; Canright 1960). Probably the carpels of the earliest angiosperms had a three-trace, three-gap system. At first the median (dorsal) trace was probably double.

A very important characteristic of primitive angiosperms is the absence of styles, the stigmas being decurrent (Fig. 10) along the margins (sutures) of the carpels (Hallier, 1901, 1912a; Parkin, 1955). Such stigmatic margins (approximated but not fused at the time of pollination, as in *Degeneria*) were the prototypes of the future stigmatic surfaces. As Kozo-Poljanski (1922, p. 121) first pointed out in his commentary on Hallier's codex of characters of the primitive angiosperm, 'the stigma developed from the sutures'. Recent investigations (Bailey and Swamy, 1951) on the morphology of primitive angiosperms have fully confirmed the origin of the stigma from such 'stigmatic crests'. Thus in the earliest angiosperms, as in some living forms, both styles and proper terminal stigmas were undeveloped (Fig. 11).

Placentation

Of the two basic types of ovule arrangement within the carpel—laminar (superficial) and sutural (submarginal)—the former is more primitive. According to many authors, the most primitive kind of laminar placentation is the scattered laminar type characteristic of *Butomaceae*, *Nymphaeaceae*, the majority of *Lardizabalaceae* and some others—see, for example, Kozo-Poljanski (1922), Zazhurilo and

Kuznetsova (1939), Joshi (1947), Gundersen (1950) and Eames (1961). However, Takhtajan (1948, 1959) proposed that lateral–laminar placentation was a more primitive type, especially as shown by such archaic angiosperms as *Degeneria*, *Drimys* sect. *Tasmannia* (Fig. 10) and certain species of *Bubbia* (Bailey and Swamy, 1951).

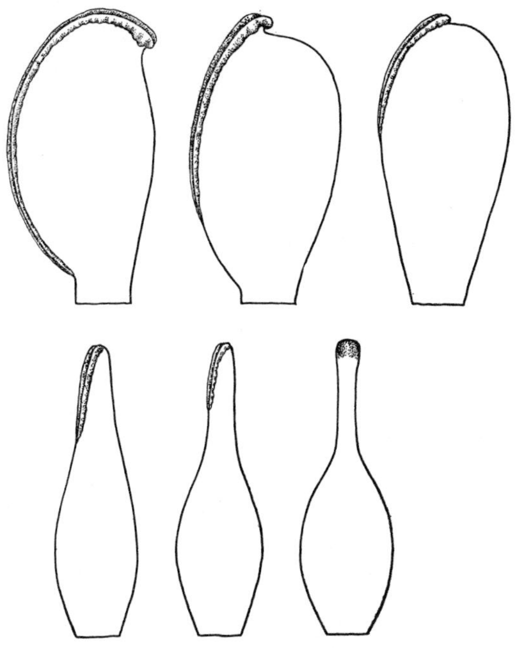

Fig. 11. Main trend of evolution in carpels from the most primitive conduplicate type. (After Takhtajan, 1954.)

Here the ovules are situated some distance from the carpel margins between the median (dorsal) and lateral veins. That the scattered type is derivative is indicated by the fact that in the family *Winteraceae* it is found only in genera like *Exospermum* which are specialised in many respects. One may also note that in the *Lardizabalaceae* the two most primitive genera *Decaisnea* and *Sinofranchetia* have ordinary sutural placentation, the ovules being situated submarginally along the margins of the carpels. Parkin (1955) was also inclined to consider the laminar placentation of the *Nymphaeaceae* or *Butomaceae* type as derivative—'a derivative rather than a primitive view is favoured for this scattered arrangement of the ovules on the carpel wall'.

Ovules

As the anatropous ovule is the basic and most widespread type amongst living angiosperms and is characteristic of the great majority of primitive families, we may postulate that the earliest angiosperms also had anatropous ovules. Although many authors, e.g. Hallier, (1912a), Kozo-Poljanski (1922, 1937) and Goebel (1933), have taken the atropous (orthotropous) ovule to be a more primitive type, there is every reason to believe that the atropous type arose from the anatropous in the course of evolution (Netolizky, 1926; Eames, 1961).

The 'double' integument is more primitive than the 'single' (Hallier, 1901, 1912a; Coulter and Chamberlain, 1903; Goebel, 1933; Joshi, 1939; Maheshwari, 1950), the former being typical of the vast majority of polypetalous dicots and most monocots, and thus the earliest angiosperms must have had bitegmic ovules. The 'outer integument' being, according to several authors, of cupular origin (developed from the 'cupule' of the seed-ferns)—see, for example, Gaussen (1946) and Walton (1953)—was probably better developed than in the present-day forms.* The ovules of primitive angiosperms are also characteristically 'crassinucellate'; that is to say, they have a massive nucellus (megasporangium) with well-developed 'parietal' (i.e. wall) tissue (Hallier, 1901, 1912a; Warming, 1913; Netolizky, 1926; Goebel, 1933, Maheshwari, 1950), and the earliest angiosperms undoubtedly had crassinucellate ovules with thick megasporangial walls and well-developed sporogenous tissue.

* Some authors consider an aril as the third integument, but as Eames (1961, p. 258) says, 'this interpretation is unfortunate, especially where it is clearly an elaboration of one, or of part of one, of the typical integuments'. In her account of the nature of arils, Komar (1965) has also come to the conclusion that 'an aril, and more so an ariloid, cannot be considered as the third integument, because the ariloid is a part of the integument, while the aril is developed on a mature and usually fertilised ovule' (p. 724).

Male and Female Gametophytes

The male and female gametophytes of angiosperms have reached, in the course of evolution, such a high degree of specialisation and simplification that the reconstruction of their ancestral type is exceedingly difficult. If our hypothesis of the neotenic origin of the angiosperms is correct, then the very earliest angiosperms must have had simplified gametophytes devoid of gametangia. One can only say it is possible that the male gametophyte was of the *Gnetum* type, with a prothallial cell, and that the female gametophyte must have been of the normal monosporic type. The seeds of the early angiosperms apparently had no period of dormancy—see Eames, 1961, p. 369.

Seed

The seeds of the early angiosperms were large (Hallier, 1901, 1912a; Corner, 1949a), with abundant endosperm and a very small and undifferentiated embryo (Pritzel, 1898; Hallier, 1912a; Martin, 1946; Wardlaw, 1955; Grushvitski, 1961). It is possible that, as in gymnosperms, the development of the proembryo in the earliest angiosperms proceeded by way of free nuclear division, without the formation of cell walls.

As the monocot embryo undoubtedly arose from the dicot (this is almost universally accepted), the most ancient angiosperms must have been dicotyledonous plants.

The basic primitive type of seed-coat (testa) in the angiosperms is the *Magnolia*-type (Fig. 12) with its strongly developed peripheral layer of parenchyma cells (Zazhurilo, 1940; Takhtajan, 1948, 1959). Endozoochory, possibly reptilian, was probably characteristic of the earliest angiosperms (Zazhurilo, 1940; Takhtajan 1948; Levina, 1957).

Fruit

The most primitive and ancestral fruit type in angiosperm evolution appears to have been a large cone-like structure consisting of large, many-seeded follicles (as in *Magnoliaceae*)—see Hallier (1910, 1912a) Bessey (1915), Gobi (1921), Kaden (1947, 1958), Takhtajan (1948, 1959) and Corner (1964). This type of fruit, developing from a multicarpellate apocarpous gynoecium, was called a 'multifolliculus' by Gobi (Figs. 12, 14/2, p. 75).

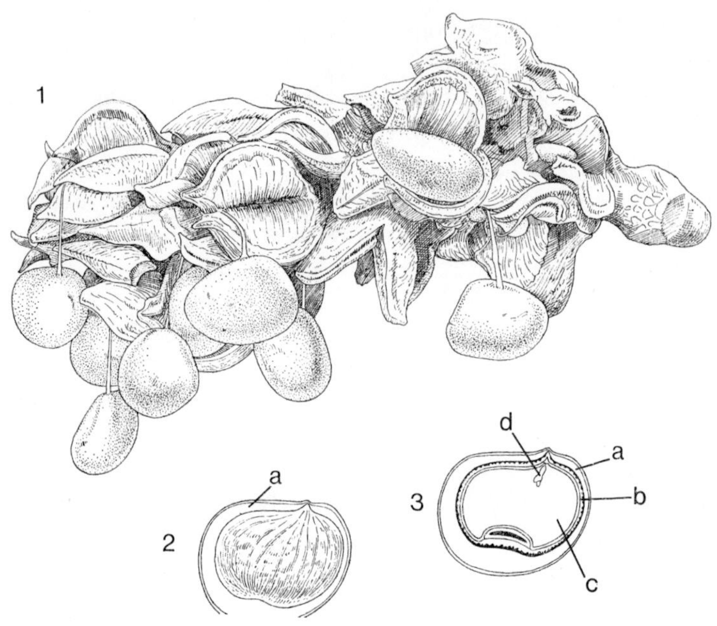

Fig. 12. Fruit and seed of *Magnolia* spp. *1*, Fruit, *2*, Seed with fleshy layer of testa removed from nearside, *3*, Seed in vertical section: *a*, fleshy layer of testa (sarcotesta); *b*, sclerotesta; *c*, endosperm; *d*, embryo. (*1* After Jackson; *2*, *3* after Prêtre.)

Karyotype

As was proposed as long ago as 1931 by Lewitsky (1931), the morphologically more primitive and basic type of chromosome of the flowering plants was one which had equally developed arms and a median or submedian centromere. Such symmetrical (or, in contemporary terminology, 'metacentric') chromosomes were those from which in course of evolution arose asymmetrical (acrocentric) chromosomes, in which the centromere is situated very close to one end (see also Stebbins, 1950).

The original karyotype of flowering plants was probably

characterised by a comparatively small number of medium-sized chromosomes. But it is very difficult to say what in fact was the ancestral basic number of chromosomes in the flowering plants. One of the main reasons for this difficulty is the inadequate knowledge of the cytology of extant primitive flowering plants, including the *Magnoliales*; but it is already clear that there is a considerable diversity of basic numbers (see Raven and Kyhos, 1965).

In the *Magnoliaceae* $2n = 38$, 76 and 114; consequently, in this family $x = 19$. In the *Degeneriaceae* and *Himantandraceae* $2n = 24$ and thus the basic number of chromosomes is different. In the *Eupomatiaceae* $2n = 20$, i.e. the basic number differs both from that of the *Magnoliaceae* and from that of the *Degeneriaceae* and *Himantandraceae*. In the *Annonaceae*, the largest family of the order and still very poorly known cytologically, $2n = 14$, 16, 18, 28 and 48, i.e. within the family there are several basic numbers, each of which is different from the basic numbers of the preceding families. In the *Canellaceae* $2n = 26$ and in the *Myristicaceae* $2n = 38$, 42 and 50 (thus in the first case $x = 19$, as in the *Magnoliaceae*). Unfortunately, in the family *Winteraceae*, in many respects very primitive and comparatively isolated within the order *Magnoliales*, cytology has been little studied. According to the latest data given by Ehrendorfer *et al.* (1968); four species of *Drimys* sect. *Tasmannia* have $2n = 26$ ($x = 13$), while in *Drimys* sect. *Drimys*, *Bubbia* and *Belliolum* $2n = 86$ and in *Zygogynum* $2n = 172$ (and thus $x = 43$, the highest basic number in the *Magnoliales*).

It appears highly probable that the basic number of chromosomes of the early flowering plants was a low one. Darlington (Darlington and Mather, 1949, p. 324) came to the conclusion that the basic number relationships of the chief families of woody flowering plants showed 7 as the common ancestral chromosome number of flowering plants. 'From this origin, 8, 9 and an increasing series have arisen on only a few occasions, whereas 14, with its diminishing series, has arisen very frequently. In this series 12 has often been stabilised, and, from its addition to 7, 19 has appeared several times.' Later, Raven and Kyhos (1965) and Ehrendorfer *et al.* (1968) reached a similar conclusion. The probability is very high that Darlington's proposal was correct and that the basic number was indeed 7. Within the limits of the order *Magnoliales*, the basic number 7 is met with in some *Annonaceae* and apparently also in Myristicaceae, i.e. in two relatively advanced families. Yet on the other hand, in such extremely primitive families as *Degeneriacea* and *Himantandraceae* $n=12$, in *Winteraceae* $x=13$ and 43, while in the very primitive family *Eupomatiaceae* $n=10$. Moreover, in the *Laurales* the large and undoubtedly very ancient family *Lauraceae* is characterised by a basic number of 12 (i.e. the same as in *Degeneriaceae* and *Himantandraceae*), while in the *Austrobaileyaceae* (the most primitive family of the *Laurales*), in some *Monimiaceae* and in the *Calycanthaceae* $x=11$. These data occasion some doubt that the basic number of chromosomes in the evolution of the karyotype of the flowering plants was in fact 7; a figure of 5, or better 6, is just as likely. Stebbins (1966) has suggested that the original number was $x=7$ or $x=6$. To be more certain, we need detailed research on the number and morphology of the chromosomes of the primitive dicotyledons.

Such, then, are the probable main features of the earliest angiosperms. Future work on primitive living forms and further development of the evolutionary morphology of higher plants may well permit us to amplify and improve our picture of the ancestral angiospermous plants, but its main outlines will probably remain without radical alteration.

CHAPTER 7

LIVING FOSSILS

A number of forms amongst the flowering plants living today is characterised by a great many extremely primitive and archaic features. They are what Darwin called 'living fossils', which through some favourable circumstance have escaped extinction and survived to the present day. They do not play any significant role in the plant world of today and usually have very restricted distributions. We have already instanced some of these primitive angiosperms in the preceding chapters. They are, for us, invaluable relics of the early stages of angiosperm evolution, and the study of them is very pertinent to the solution of many fundamental questions concerning the morphology and phylogeny of the angiosperms.

MAGNOLIALES

The most primitive families of living angiosperms are contained within the order *Magnoliales*. Different authors have varied in their circumscriptions of this group. Here we include in the order the families *Winteraceae*, *Magnoliaceae*, *Degeneriaceae*, *Himantandraceae*, *Eupomatiaceae*, *Annonaceae*, *Canellaceae* and *Myristicaceae*. Many more ancient and primitive features are to be found in these eight families than in any other order of angiosperms. They include vesselless wood, primitive flowers with spirally arranged free parts on an elongated axis, broad laminar stamens undifferentiated into filament and connective, incompletely closed carpels, and monocolpate pollen of ancestral gymnosperm type. These primitive features are always found in association with more specialised ones; each family is specialised in its own way. It is therefore very difficult to say which is the most primitive. Some authors, e.g. Hutchinson (1959a), consider the *Magnoliaceae* to be the most primitive, others, e.g. Gundersen (1950), the *Winteraceae*; it is immaterial for our purposes with which we begin here.

Winteraceae

The family *Winteraceae*, consisting of seven genera and about 120 species, occurs in Central and South America (Mexico southwards to the Straits of Magellan), Juan Fernandez, the Malay Archipelago (except Sumatra, Java and Timor), New Guinea, eastern Australia, Tasmania, Lord Howe Island, Fiji, New Zealand, New Caledonia and the Solomon Islands (Smith, 1945a), with one species in Madagascar (*Bubbia perrieri* Capuron). The distribution of this family around the Pacific Basin is thus clearly apparent (Fig. 29, p. 146). Fossil remains of *Drimys* are known from the west coast of North America, from eastern Australia and from Antarctica (Graham Land). The family has been comparatively well studied morphologically, e.g. Van Tieghem (1900), Bailey (1944b), Bailey and Nast (1943a, 1943b, 1945a), Nast (1944), Hotchkiss (1955b), Tucker (1959), Sampson (1963), Bhandari (1963), Tucker and Gifford (1964, 1966), and others. It consists of trees and shrubs with entire pinnately veined aromatic exstipulate leaves. The nodes are trilacunar and the xylem without vessels. This absence of vessels in *Winteraceae* is a survival from ancient gymnospermous days, and is without doubt one of the family's most notable features. The tracheids are long and thick-walled, usually with two to three rows of large circular bordered pits, though in the early (spring) wood of some species of *Drimys* and *Bubbia* the more primitive type with scalariform bordered pits occurs. The rays are heterogeneous and the wood parenchyma is normally diffuse.

In flower structure the *Winteraceae* are in some respects primitive (sometimes extremely so) and in other respects advanced, (even markedly so, at times). Thus the flowers may be bisexual or unisexual, and the carpels primitive or specialised. Unlike the *Magnoliaceae* the *Winteraceae* have a short floral axis, and the perianth is differentiated into distinct calyx and corolla (Fig. 13). The androecium of *Winteraceae* consists of (5–6)–12–many stamens arranged in 2–5 turns of a low spiral. The stamens of *Belliolum* are comparatively primitive and have a protruding connective, but the stamens of the other genera show stages of specialisation. The pollen grains are comparatively advanced—they are united in permanent tetrads and have a distal aperture reduced to a circular pore and have a reticulate exine. The carpels are comparatively few (rarely as many as 24, and sometimes reduced to one), usually in one series, and although mostly free are sometimes more or less united, especially in fruit (tribe *Exospermeae*). Nevertheless, in members of *Drimys* sect. *Tasmannia* occurring in the Old World we find very primitive carpels reminiscent of folded (conduplicate) young leaves (Fig. 10, p. 61). Moreover, in certain species of this

Fig. 13. Flowering branch of *Drimys piperita*: *1*, general view; *2*, flower; *3*, carpel; *4*, fruits.

section, e.g. *Drimys piperita*, the carpel margins at pollination time are only approximated, not fused, and are united not by the epidermes but solely by the papillose hairs of the stigmatic surfaces. The latter form broad zones running along the inner surfaces of the carpel and extend laterally from the extreme margins to the ovuliferous zones. Except in its papillose hairs, this primitive stigmatic surface bears little resemblance to the normal strictly localised stigma, which forms a distinct part of the carpel. Pollen grains fall and germinate on the papillose hairs covering the free margins of the carpels, and the pollen tubes penetrate between the interlocking papillose hairs of the closely-adhering inner stigmatic surfaces of the carpel. Strictly speaking, therefore, only the papillate surfaces of the extreme margins of the carpels function as a stigmatic surface, the inner papillate zones performing the function of the so-called 'transmitting tissue' in facilitating the passage of the pollen tubes. A similar type of carpel is found in some species of *Bubbia* and *Exospermum*.

The primitive carpel structure of *Winteraceae* is paralleled by the primitive position of the ovules. As we have seen, the papillate marginal zone of the inner surface of the carpels serves both as a stigma (the outer recurved part) and as transmitting tissue (the inner part). The ovules are therefore at some distance from the carpel margins, situated between the median and lateral veins. In *Drimys piperita* and related species, and in some species of *Bubbia*, the ovules occupy the lateral part of the inner surface of the carpels. The association in these forms of a similar type of placentation with the most primitive type of carpel makes possible the consideration of this *Drimys*-type of placentation as the basic ancestral type in living angiosperms (Fig. 10). Since in the course of evolution the carpel margins fused and the stigmatic surfaces gradually became localised at the distal ends of the carpels, the broad sterile papillate marginal zones gradually shortened and vanished, for they were no longer necessary in the closed carpel with its specialised stigmatic surface. In this way the ovules 'shifted', as it were, closer to the carpel margins, although they never achieved a strictly marginal position. As a result, the ordinary sutural placentation developed from the primitive lateral-laminar placentation of the *Drimys*-type. The flowers are entomophilous or rarely anemophilous (as in *Pseudowintera*—see Sampson, 1963).

Magnoliaceae

This family of 12 genera and about 230 species occurs mainly in the subtropics of the northern hemisphere. In the Late Cretaceous and in the Tertiary the *Magnoliaceae* were very widely distributed in northern Eurasia and North America, growing not only in Europe and the Caucasus but even in Greenland, Spitzbergen and Alaska. Their present distribution, however, is very discontinuous (Fig. 28, p. 144). They are concentrated mostly in the two largest refugia of ancient woody Tertiary vegetation: east and south-east Asia; south-east United States, Central America and the Greater Antilles. The greatest specific diversity is concentrated in these two regions, and the greatest number of species is found in the eastern Himalayas and south-west China and Indo-China. Only a few *Magnoliaceae* are found in the southern hemisphere—in Brazil and in the Malay Archipelago.

The *Magnoliaceae* are deciduous or evergreen trees and shrubs with alternate, entire (or rarely lobed) pinnately-veined leaves. They have large deciduous stipules which enclose the young buds. Anatomically they are mostly of rather primitive type, their vessels usually having scalariform perforation; in *Talauma* and *Kmeria* the number of perforation bars sometimes reaches 25, in *Magnolia pterocarpa* 20, and in *Manglietia* 30 to 40 (Ozenda, 1949). But even within *Magnolia* itself we can see all transition stages, from primitive vessels with scalariform perforation to specialised vessels with simple perforation. The primitive character of the vessels in *Magnoliaceae* is also shown by the usually scalariform intervascular pitting (though opposite in *Magnolia fraseri* and *Liriodendron*). This shows that the vessel elements arose from tracheids with scalariform and not circular bordered pits. The wood parenchyma in *Magnoliaceae* is frequently of the terminal apotracheal type, and the rays are usually of the rather primitive heterogeneous type; only rarely (in a few species of *Magnolia*—*M. acuminata*, *M. kobus* and *M. obovata*) are they homogeneous (Canright, 1955). The nodes are multilacunar, with 6 to 17 leaf-traces (Canright, 1955).

The flowers of *Magnoliaceae* are even more primitive than their wood. They are rather large, reminiscent in aspect of some large bennettitalean strobili, solitary* and terminal or more rarely (*Michelia*, *Tsoongiodendron* and *Elmerrillia*) axillary, and bisexual or rarely (*Kmeria*) unisexual (Fig. 14). A rather elongated floral axis (receptacle) is frequently characteristic; it has all the structural peculiarities of a stem, but of a shortened stem with shortened internodes. It is interesting to note that, as Skvortsova (1958) showed, the receptacle of *Magnolia* is without vessels. The perianth consists of calyx and corolla, often very similar but in some species of *Magnolia* and *Michelia* quite distinct. In this family one can follow a reduction in the number of perianth parts from 12, 18 or more (*Aromadendron*) to six (*Kmeria* and some species of *Michelia*). The flowers of *Magnoliaceae* are entomophilous.

The stamens in *Magnoliaceae* are numerous, free and spirally arranged (a primitive feature), and also rather primitive in structure, being trinerved, more or less laminar and in most cases produced above the microsporangia, usually undifferentiated into filament and connective, and bearing two pairs of linear microsporangia usually deeply immersed in sterile tissue (Canright, 1952). But as well as the broad primitive stamens of *Elmerrillia* spp. *Talauma gigantifolia*, *T. mexicana*, *Manglietia forrestii*, *Magnolia maingayi*, *M. stellata*, etc., there are also stamens specialised in one way or another. For example, in many typical members of the family, such as *Liriodendron* spp., *Magnolia campbellii*, *M. watsonii*, *M. obovata*, *Michelia yunnanensis* and *M. fuscata*, the stamens are more or less differentiated and, whilst they do not have a proper 'filament', the sterile part is somewhat narrowed. In the specialised stamens of *Liriodendron* the connective is not produced beyond the microsporangia. The number of vascular bundles has also been reduced in the course of evolution; in some species, like *Manglietia hookeri* and *Talauma singapurensis*, the stamens have five to seven traces; sometimes (as in other species of *Manglietia* and *Talauma*) the number of traces varies from three to seven; in the majority of *Magnoliaceae* there are three traces, and in a few there is only one. Stamens with one trace are

* According to Eames (1961), 'In the presence of occasional accessory flowers in several genera and of clusters of flowers ("axillary") in *Michelia*, there is evidence that the usually solitary flowers represent reduction from an inflorescence' (p. 389).

Fig. 14. *Manglietia insignis*: *1*, flower; *2*, fruiting branch; *3*, stamen, (*a*) ventral and (*b*) dorsal views; *4*, stamen in cross-section, adaxial surface facing downwards; *5*, carpel; *6*, carpel in section, showing ovules.

most usual in the genus *Michelia*, where they are characteristic of the majority of species. Also in course of evolution immersed microsporangia have become 'protuberant', rising to the surface of the stamen, as in *Magnolia denudata* and *Liriodendron*. Finally, as the stamens have narrowed, the microsporangia have come to occupy the 'marginal' position and open extrorsely, as in *Liriodendron* and species of *Michelia* (Canright, 1952; Ozenda, 1952; Eames, 1961).

The pollen in the *Magnoliales* is of a very primitive monocolpate type (Fig. 9, p. 60). In the dry state the pollen grains become boat-shaped due to the invagination of the comparatively thin membrane of the furrow. The sporoderm in *Magnoliaceae* is very similar to types found in the *Bennettitales*, cycads and ginkgo (Wodehouse, 1935; Takhtajan, 1947; Canright, 1953), an indication of its primitiveness. The largest and evidently most primitive pollen grains are characteristic of *Magnolia* and *Talauma*, the smallest of *Michelia* and *Elmerrillia* (Canright, 1953).

In the flowers of *Magnoliaceae* the carpels are usually numerous, as in *Michelia*, *Tsoongiodendron* and *Magnolia*, more rarely few (five to six in *Aromadendron* and *Kmeria* and two to eight in *Pachylarnax*); they are spirally arranged, free or rarely more or less united, in *Pachylarnax* (of Assam, Indo-China and the Malayan Peninsula) forming a loculicidal capsule. Thus the change from apocarpy to syncarpy has taken place even within so primitive a family as the *Magnoliaceae*. There is also interesting variation in the number of ovules. The greatest number is found in the primitive genus *Manglietia* and in *Tsoongiodendron*. On the other hand, in the primitive genera *Magnolia* and *Talauma* there are only two ovules in each carpel. In all the genera (except *Elmerrillia* of New Guinea and the Malayan Archipelago) the carpels have a more or less distinct style. Finally, *Elmerrillia* and *Manglietia* share another primitive feature in that the margins of the conduplicate carpels are only slightly fused. In *Michelia* the margins of the conduplicate carpels are fused at anthesis only in the basal half, while in the remaining genera the margins are fused along their entire length (Canright, 1960). On the basis of all the available data, Canright considered the carpels of *Elmerrillia* and *Manglietia* to have the most primitive structure in the *Magnoliaceae*, and those of *Liriodendron* the most specialised.

On the whole, in contrast to their stamens, the carpels in this family show a fairly high degree of evolutionary advancement.

The fruit of the *Magnoliaceae* is usually a spiral multifolliculus—the most primitive known type of fruit (Fig. 12, p. 66).

Another primitive feature of the family is the very small embryo in the seeds and the presence of abundant endosperm, a characteristic of the *Magnoliales* in general (Fig. 12). The seed-coat (testa) is also primitive in structure; as Zazhurilo (1940) showed, the *Magnolia* type of testa (Fig. 12) with its characteristic external somewhat fleshy layer (sarcotesta) is primitive, but the testa of *Liriodendron*, which is without sarcotesta, shows specialised features; see also Takhtajan (1959).

Finally, the basic chromosome number in the family is 19 (Whitaker, 1933; Yasui, 1937; Janaki Ammal, 1953; Canright, 1953). *Magnolia*, which is cytologically the best-known genus, exhibits a distinct polyploid series ($2n = 38$, 76 and 114).

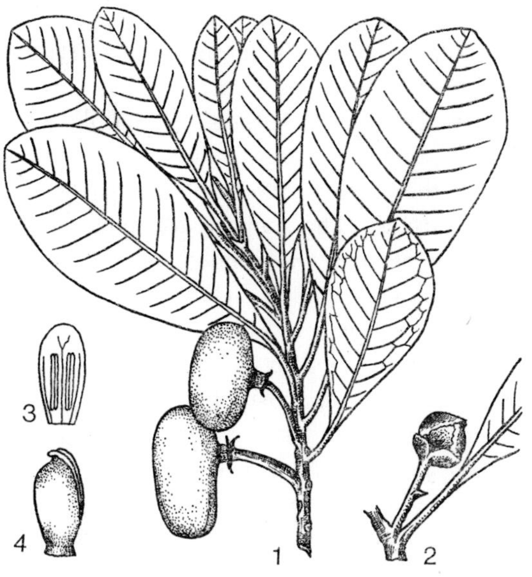

Fig. 15. *Degeneria vitiensis*: *1*, fruiting branch; *2*, part of a flowering branch with bud; *3*, stamen; *4*, carpel. (After Takhtajan, 1961.)

Degeneriaceae

This family, which is endemic in the Fiji Islands, stands very close to the *Magnoliaceae* and consists of one monotypic genus *Degeneria* (Fig. 15). Morphologically *Degeneria* has many very interesting peculiarities (Bailey and Smith, 1942; Swamy, 1949; Lemesle and Duchaigne, 1955; Eames, 1961; Dahl and Rowley, 1965; Raven and Kyhos 1965).

The leaves of *Degeneria* are entire and pinnately veined; but unlike those of the *Magnoliaceae*, they are without stipules, and arise from pentalacunar nodes. The wood structure is very primitive; the vessel elements are thin-walled, angular in cross-section, with a scalariform perforation with numerous perforation-bars and scalariform intervascular pitting; the rays are heterogeneous.

The flowers are solitary, long-pedicellate, and supra-axillary. The presence on the pedicel of two or three bracts has been taken as an indication that in the past there were lateral flowers (Eames, 1961).

In contrast to the *Magnoliaceae*, the floral axis of *Degeneria* is short and its vascularisation much simpler. The perianth is cyclic and clearly differentiated into calyx and corolla, but although the petals are considerably larger than the sepals and different in form, they are similar to them in their anatomical structure. But the greatest interest of the flowers of *Degeneria* lies in their very primitive stamens and carpels. The stamens are numerous, broad and undifferentiated into filament and connective; they have a distinct median vein, which dichotomises at the apex, and two lateral veins. Two pairs of long narrow microsporangia are situated one on each side between the median and lateral veins on the abaxial surface of the stamen; as in most of the *Magnoliaceae* they are immersed in sterile sporophyll tissue. Thus the *Degeneria* stamen (Fig. 8, p. 58) is extremely primitive in construction and may well be regarded as the most primitive angiosperm stamen. The pollen likewise is very primitive, being similar in form and structure to the magnoliaceous type. It is the carpel of *Degeneria*, however, that is its most notable feature. In contrast to the multicarpellate *Magnoliaceae* the number of carpels is reduced to one; but this one remaining carpel shows an extremely primitive conduplicately folded structure. The carpel margins are not only completely free, but before anthesis are noticeably distant from each other. There are numerous ovules in two rows situated quite remote from the margins. At flowering time the broad areas between the carpel margins and the ovules stand close together but are not actually coherent except in the lower part of the carpel; the stigmatic surfaces extend along the margins of the carpel on the inner sides, each forming a zone between the margin and the ovuliferous region, and are thus of a very archaic type. Only when the fruit develops do the contiguous adaxial surfaces become concrescent. Pollen grains are caught by the outcurving glandular-hairy carpel margins, where they germinate, the pollen tubes growing down between the loosely interlocking papillose, glandular hairs of the marginal areas to reach the ovules.

Himantandraceae

The primitive family *Himantandraceae* is close to the *Magnoliaceae* and *Degeneriaceae* and consists of one genus *Himantandra* of three species occurring in north-east Australia, New Guinea and the Moluccas. In some ways it is a link between the *Magnoliaceae* and *Degeneriaceae* on one hand and the *Annonaceae* on the other, though phylogenetically it represents a lateral branch. In *Himantandra* there are not only some specialised features, but also many extremely archaic ones (Diels, 1919; McLaughlin, 1933; Ozenda, 1952; Bailey, Nast and Smith, 1943; Lemesle, 1955; Buchheim, 1962). The species of *Himantandra* are covered with peculiar peltate scales—a specialised feature. The wood structure is on the whole rather specialised; the vessels of the mature wood usually have simple perforation (in young stems both simple and scalariform perforation occur) and alternate intervascular pitting. The wood parenchyma occurs in concentric circles and the rays are heterogeneous. Unlike *Magnoliaceae* and *Degeneriaceae*, *Himantandra* has trilacunar nodes. The leaves are alternate, entire, pinnately veined and, as in *Degeneria*, without stipules. The flowers are usually solitary and terminal on short axillary branches (Fig. 16), but these are sometimes accompanied by one or two lateral flowers. Their presence indicates that here, as in *Degeneria*, we have a reduced inflorescence. The

Fig. 16. Flowering branch of *Himantandra balgraveana*.

perianth consists of calyx and corolla; the former is of very peculiar construction, consisting of two coriaceous deciduous sepals which form a calyptra which protects the flower in the bud stage. The corolla consists of a few lanceolate petals and is followed by petaloid staminodes. There are numerous stamens of the *Degeneria* type, but markedly longer and narrower and with shorter microsporangia. The pollen grains are like those of *Degeneria*, but smaller. Unlike in *Degeneria*, staminodes also occur *above* the stamens. Six to ten spirally arranged conduplicate carpels make up the slightly syncarpous gynoecium; the carpels are clearly differentiated into a fertile part and a style which bears the decurrent stigma, and each contains one, or sometimes two, pendulous ovules. They are thus considerably more specialised than those of *Degeneria*; in the flower they are slightly connate at the base, but in the fruit they are more fully united. The seeds have a small embryo and an oily endosperm.

Eupomatiaceae

The highly interesting family *Eupomatiaceae* consists of two species of the genus *Eupomatia* found in east Australia and New Guinea. Like other members of the order *Magnoliales*, the *Eupomatiaceae* exhibit both very primitive and specialised features (Lemesle, 1938; Hotchkiss, 1958; Uphof, 1959a; Eames, 1961). They are trees and shrubs with entire pinnately veined exstipulate leaves; the nodes are multilacunar, with 7–11 traces. The wood is very primitive in structure; it has very long (1000 μ), narrow (60–70 μ) vessel elements with very oblique end walls, scalariform perforation with numerous (20–150, or more) perforation bars, and scalariform or opposite pitting on the lateral walls. The openings of the scalariform perforation differ from the scalariform bordered pits of the lateral walls only in the absence of a pit membrane. The rays are heterogeneous. The fibres have simple or bordered pits, and the sieve elements are long and narrow, with gradually tapering ends. The flowers of *Eupomatia* (Fig. 17) are rather large, terminal, and solitary; in the young stage each flower is protected by a deciduous cap or calyptra which is attached to the margin of the expanded concave receptacle. Some authors, e.g. Endlicher (1841), Gun-

Fig. 17. Flowering branch of *Eupomatia laurina*.

dersen (1950), Hutchinson (1959a) and Uphof (1959a), regard the calyptra as a modified perianth in which the sepals and petals are united into one structure, but the majority—among them Baillon (1868-1870), Diels (1912) and Eames (1961)—regard the flower of *Eupomatia* as apetalous and the calyptra as a modified bract. The stamens are numerous and inserted on the rim of the hollow receptacle. They are comparatively primitive, laminar, with three to seven traces, and a connective produced beyond the microsporangia. The special peculiarity of *Eupomatia* is that the *inner* stamens are modified into petaloid staminodes, the outer stamens being fertile (contrast *Himantandra*, in which, as we have seen, staminodes occur both above and below the stamens). No less peculiar is the structure of the syncarpous gynoecium, which consists of numerous spirally arranged carpels inserted on the concave surface of the turbinate receptacle. The carpels have usually been regarded as free and immersed in the receptacle, as in *Nelumbo*, but, as the work of Hotchkiss and Eames (Eames, 1961) has shown, the spirally arranged carpels are united (like those of *Zygogynum* in the *Winteraceae*). In spite of their connation, the carpels are slightly open, devoid of a style, and characterised by decurrent marginal stigmatic surfaces similar to those of *Degeneria* and *Drimys* sect. *Tasmannia*; they have several traces. The pollen grains have two or three distal furrows. Pollination is effected by beetles, and the flowers are adapted for pollination by only one species of beetle. The presence of thick fleshy staminodes is associated with this most elaborate cantharophily. The seeds of *Eupomatia* have a well-developed ruminate (wrinkled) endosperm and a very small embryo, a primitive feature. The family is somewhat isolated within the *Magnoliales*; it is closest to the *Himantandraceae*, according to Uphof (1959a), but the basic number of chromosomes in these two families is not the same ($n=12$ in *Himantandra* and 10 in *Eupomatia*; see Raven and Kyhos, 1965).

The families *Annonaceae*, *Canellaceae* and *Myristicaceae* are also of great interest.

Annonaceae

The large family *Annonaceae* contains about 120 genera and 2100 species occurring in the tropical and, to a lesser extent, subtropical regions of Asia, Africa, America and Australasia, with the greatest number of species in the Old World. This family has attracted the attention of many workers, especially Hutchinson (1923), Diels (1932), Asana and Adatia (1947), Corner (1949b), Van der Wyk and Canright (1956), Fries (1959) and Canright (1963). It comprises trees, shrubs or lianas with simple, entire, pinnately veined exstipulate leaves. Anatomically the wood is rather specialised, the vessel elements of the majority of genera being comparatively short, with almost transverse end walls and exclusively simple perforation. The intervascular pitting is predominately alternate, the rays are homogeneous or weakly heterogeneous, and the fibres are comparatively short, thick-walled and mostly libriform. The flowers are large, bisexual or rarely unisexual, and with a cyclic perianth of three whorls (an outer of three, or rarely two, sepals and two inner of three petals each). The stamens are usually numerous, spirally arranged, and with broad thick connectives; often they are undifferentiated into filament and connective, but laminar and produced above the microsporangia, as in species of *Polyalthia*. In some genera the microsporangia are completely sunken, as for example in *Polyalthia* and *Fissistigma*. Each stamen is provided with one trace. The pollen is monocolpate. Pollination is apparently carried out mostly by beetles. The gynoecium is apocarpous, or rarely (in the African genera *Monodora* and *Isolona*) paracarpous, and the number of ovules per carpel varies from many to one; the seeds have a copious, ruminate endosperm and small embryo. The *Annonaceae* are related to the *Magnoliaceae*, but noticeably more advanced.

Canellaceae

The *Canellaceae*, composing five genera and about seventeen species, occur in tropical America, east tropical Africa and Madagascar. They are mostly small evergreen trees with simple entire pinnately veined exstipulate leaves and trilacunar nodes. The wood structure is on the whole more primitive than in the *Annonaceae*. The vessel elements are usually very long, with oblique end walls and scalariform perforation plates; in most of the genera there are 10–20 bars in the perforation plate, but in *Cinnamodendron* there are usually 50–100. Wood parenchyma is diffuse apotracheal, or apotracheal and sparingly paratracheal, rarely (in *Canella* and *Pleodendron*) predominantly vasicentric (a specialised feature). The rays are mostly homocellular, but some heterocellular rays are also present, and the fibres have comparatively large bordered pits (Metcalfe and Chalk, 1950; Lemesle, 1955; Wilson, 1960, 1965). The flowers are bisexual, with three thick imbricate sepals and 4–12 petals, which are free or connate at the base, or rarely (in the Madagascan genus *Cinnamosma*) connate to half-way and forming a tube. The stamens number 7–12, or rarely 20–40, and are united into a tube. The pollen is monocolpate. The gynoecium of two to six carpels is paracarpous with a single thick style. The fruit is a berry with two or more seeds. The seeds have a small embryo and abundant and sometimes ruminate (*Cinnamosma*) endosperm. (For more detailed accounts of this family, see the works of Occhioni, 1948; Wilson, 1960, 1965; and Parameswaran, 1962.) It shows affinity with the *Annonaceae* and the *Myristicaceae* and possibly had a common origin with them.

Myristicaceae

This family of 16 genera and about 400 species is evidently close to the two preceding families and occurs in the tropical regions of Asia, Africa and Madagascar and America. The *Myristicaceae* are noticeably more primitive than the *Annonaceae* in their wood structure, but are more advanced florally; the flowers are apetalous and unisexual, with the stamen filaments united into a column, and a monomerous gynoecium. (Further details of this family may be found in the works of Garratt, 1933; Smith and Wodehouse, 1938; Sastri, 1954, 1959; Joshi 1946; Van der Wyk and Canright 1956; Nair and Bahl, 1956; Uphof, 1959b; Wilson and Maculans, 1967.)

LAURALES

The order *Laurales*, which follows, is considerably more specialised than the *Magnoliales* but so closely connected with it that in the majority of systems it is not separated as distinct. The *Laurales* are very old and possibly arose from the most ancient vesselless

Fig. 18. *Sarcandra glabra* (*Chloranthaceae*): *1*, habit; *2*, flower.

Magnoliales. They are all woody plants, except the parasitic genus *Cassytha* (*Lauraceae*) and most species of *Chloranthus*. There are only two vesselless genera—*Amborella* and *Sarcandra* (Fig. 18)—amongst them, and in all except *Austrobaileya*, *Chloranthaceae* and *Lactoris* (Fig. 19) the leaves are devoid of stipules. The nodes are

Fig. 19. *Lactoris fernandeziana*: *1*, fruiting branches; *2*, flower; *3*, fruit; *4*, stamen, (*a*) ventral and (*b*) dorsal views; *5*, carpel, opened to show ovules.

usually unilacunar with two discrete traces. A specialised wood structure is characteristic of the majority. As in the *Magnoliales*, there are usually secretory cells in the parenchyma. The flowers are usually cyclic. The pollen is monocolpate or of a derivative type. In the majority of the families the gynoecium is apocarpous, The nature and degree of development of the endosperm vary considerably—sometimes it is absent. The embryo is often very small.

Austrobaileyaceae

One of the most primitive representatives of the order *Laurales* is the small Australian family *Austrobaileyaceae* (Croizat, 1940; Bailey and Swamy, 1949). It consists of the genus *Austrobaileya* (Plate XII), with two species which occur in Australia (Queensland). It is a large climbing shrub with opposite or subopposite, entire, pinnately veined coriaceous leaves with small deciduous stipules. The nodes are unilacunar with two leaf traces, and, as we have seen in the last chapter, this type of node is considered by some contemporary authors to be the most primitive amongst the angiosperms. The wood shows both primitive and specialised features. The vessel elements are 30–200μ in diameter, and in the narrower elements the end walls are very oblique; the perforation is always scalariform. The wood parenchyma is paratracheal, i.e. of a specialised type. The sieve elements are very primitive in structure being more like the sieve cells of gymnosperms than the sieve tubes of angiosperms. The flowers are solitary, axillary, bisexual, and with spirally arranged parts; the perianth consists of sepals and petals connected by transitional members. The stamens are of a primitive type, broad, laminar, and with long microsporangia. Sepals, petals and stamens all have two vascular bundles. The pollen is monocolpate. The gynoecium is apocarpous, of about eight carpels; the latter have a two-lobed style with a decurrent stigma. The median carpel trace is double at the base. The fruitlets are berries (L. S. Smith, personal communication). Thus, in spite of its rather specialised (lianous) life-form, *Austrobaileya* exhibits a series of very primitive features in both its vegetative and reproductive organs. Evidently it stands close to the ancestral lauralean type, although it is undoubtedly an evolutionary side-branch; it is closest to the *Amborellaceae* and *Monimiaceae*.

Amborellaceae

The *Amborellaceae* consist of one monotypic genus, *Amborella*, endemic in New Caledonia, an island well-known for its many peculiar and archaic forms. *Amborella* is a shrub with entire or lobed leaves, unilacunar nodes and very primitive vesselless wood. The tracheids are very long, with circular or sometimes scalariform bordered pits. The rays are heterogeneous. The flowers are unisexual, but show many primitive features—a perianth of spirally arranged members which pass downwards into the bracteoles, numerous stamens with broad filaments, and an apocarpous gynoecium (Bailey and Swamy, 1948, 1949; Bailey, 1957).

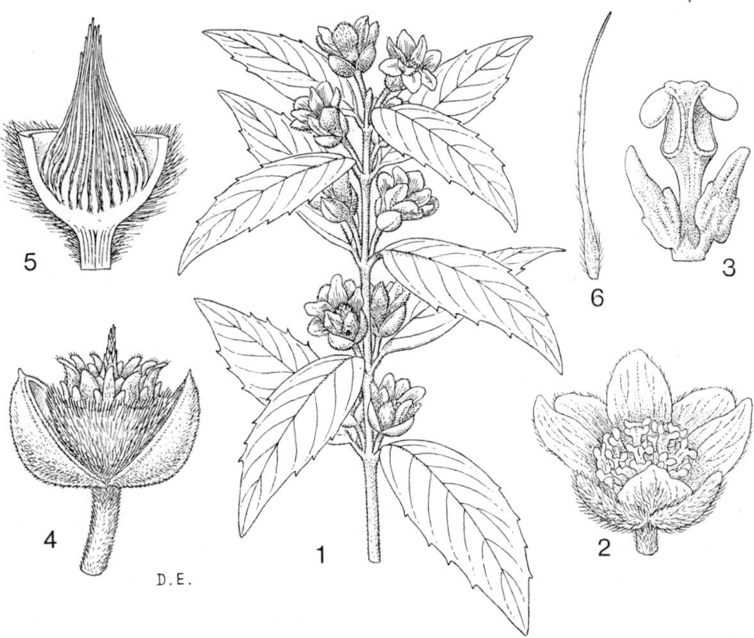

Fig. 20. *Atherosperma moschatum* (*Monimiaceae*): *1*, flowering branches (male); *2*, male flower; *3*, stamen; *4*, female flower with perianth removed; *5*, receptacle and carpels in longitudinal section; *6*, carpel.

Monimiaceae

The family *Monimiaceae* (Fig. 20) consists of about 32 genera and 450 species which are widely distributed in the tropics and subtropics, mostly in the southern hemisphere. Like all the preceding families, the *Monimiaceae* possess both primitive and advanced features (Garratt, 1939; Money, Bailey and Swamy, 1950; Erdtman, 1952; Lemesle and Pichard, 1954; Stern, 1954). In wood structure the family is rather diverse; the vessel elements are 300–500 μ in length, with slightly to very oblique end walls, rarely simple but usually scalariform perforation with ten or more bars, and scalariform to alternate pitting on the lateral walls. The rays are weakly to distinctly heterogeneous, and the wood parenchyma mostly apotracheal. The fibres may have simple or distinctly bordered pits. Also very varied is the flower structure; commonly there is a more or less urceolate receptacle which bears on its inner surface usually numerous stamens and free carpels.

Calycanthaceae

The small family *Calycanthaceae* is also of this affinity; it consists of two small genera which occur in China (*Chimonanthus*) and in the south-eastern states of North America (*Calycanthus*). They are evergreen or deciduous shrubs with opposite, entire, pinnately-veined exstipulate leaves and unilacunar bifascicular nodes. In contrast to the *Monimiaceae*, the vessels have exclusively simple perforation. The wood parenchyma is paratracheal, and the sieve tubes are also of a specialised type. The flowers are solitary and terminal on leafy lateral branches, bisexual, spiral, with a concave receptacle, numerous stamens, and numerous free carpels situated on the inner surface of the urceolate receptacle. The perianth consists of numerous spirally arranged petaloid members undifferentiated into calyx and corolla. The stamens have a broad connective produced above the anthers, and the pollen grains have two distal furrows. The carpels have one or two ovules, a long filiform style (a specialised feature) and decurrent stigmas (a primitive feature). Pollination is carried out by beetles, which are attracted by special food bodies found at the apices of the connectives. The fruitlets are dry and indehiscent (i.e. the fruit is a multinucula as in *Ranunculus*) and the seeds are without endosperm (Diels; 1916; Daumann, 1930; Grant, 1950b; Schaeppi, 1953; Fahn and Bailey, 1957; Cheadle and Esau, 1958; Eames, 1961, Nicely, 1965).

The evolution of the *Laurales* culminates in the specialised families *Lauraceae* and *Gyrocarpaceae*, which are rather far removed from such primitive members of the order as the *Austrobaileyaceae* and *Amborellaceae*.

Trochodendrales

Trochodendraceae and Tetracentraceae

A rather isolated position among living angiosperms is occupied by the extremely interesting order *Trochodendrales*, which consists of the two monotypic eastern Asiatic families *Trochodendraceae* (Fig. 21) and *Tetracentraceae* (Fig. 22, p. 93). These two families have been fairly well studied morphologically, e.g. by Van Tieghem (1900), Wagner (1903), Bailey and Thompson (1918), Bailey and Nast (1945b), Smith (1945b), Nast and Bailey (1945), Croizat (1947b), Bondeson (1952), Yoffe (1962, 1965), Pervukhina (1962), and Pervukhina and Yoffe (1962). The chief distinguishing feature between them and the *Magnoliales* and *Laurales* is their different type of pollen grain; a distal furrow is absent and is replaced by three meridianal furrows (tricolpate pollen). The latter arrangement of apertures (Fig. 21/4), so characteristic of the dicots, is found only in angiosperms and is unknown in the gymnosperms. This tricolpate type of grain evidently arose in the course of evolution from a more primitive type with one polar furrow. In the *Trochodendrales* the meridianal furrows are still comparatively primitive. The outstanding feature of the order is its vesselless wood (Fig. 4, p. 52), which is hardly in keeping with its tricolpate pollen and comparatively highly specialised apetalous flowers.

Trochodendron aralioides, a native of South Korea, Japan and Taiwan (Formosa), is a small evergreen tree with entire pinnately-veined exstipulate leaves and very primitive wood. The tracheids are very long, with typical scalariform bordered pits in the early wood, though in the late wood, when the tracheids are rather narrower, the pits assume a rounded outline. In spite of this archaic 'drimysian' wood the flowers of *Trochodendron* are of a comparatively specialised type. They are small and aggregated into terminal cymose inflorescences (racemiform pleiochasia) and

Fig. 21. *Trochondendron aralioides*: *1*, flowering branch; *2*, flower; *3*, carpel in section; *4*, pollen grain; *5*, immature fruit; *6*, dehisced fruit. (After Takhtajan, 1961.)

completely devoid of a perianth (Fig. 21), though furnished with two small bracteoles which have sometimes been mistaken for the remnants of a calyx. They are bisexual,* with about 70 stamens arranged spirally and adnate by their filaments to the lower parts of the carpels. They are of a rather specialised type clearly differentiated into filament and connective. The gynoecium consists of five

* According to Keng (1959), *Trochodendron aralioides* appears to be androdioecious.

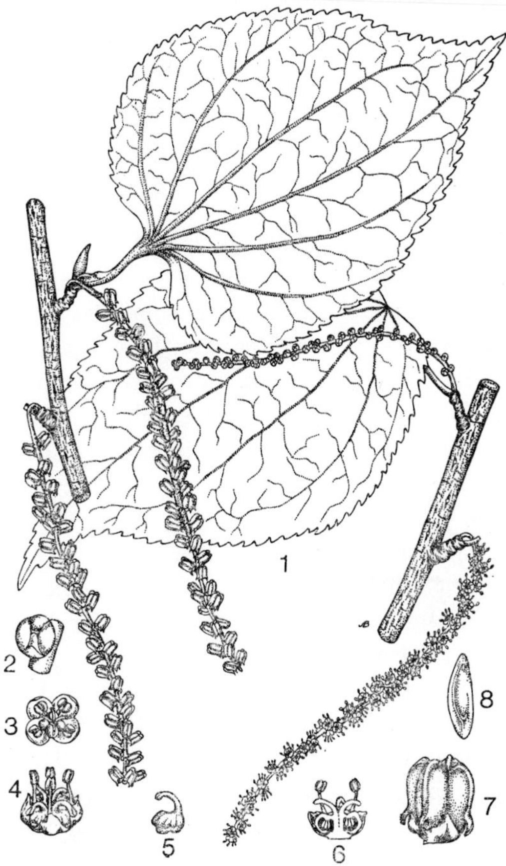

Fig. 22. *Tetracentron sinense*: *1*, flowering branches; *2*. bud; *3*, *4*, flowers at various stages of development; *5*, a single carpel; *6*, carpels in section; *7*, fruit; *8*, seed. (After Takhtajan, 1961.)

to 11 (sometimes more) carpels united laterally, the lateral bundles of neighbouring carpels also being fused. The connation is not total, however, and a considerable part of each carpel remains free. The styles are rather primitive, having decurrent stigmatic surfaces. On the convex dorsal parts of the carpels nectar is abundantly secreted, showing that *Trochodendron* is entomophilous, at least partly, but the entomophily is probably of secondary origin.

In each carpel there are 25–30 ovules. The seeds have a very small embryo and abundant endosperm. Thus in *Trochodendron* we observe a mixture of primitive and advanced characters.

The monotypic genus *Tetracentron* is found in central and south-west China, upper Burma, N.E.F.A., Bhutan and east Nepal. *Tetracentron* is a medium-sized tree (as I have seen it in Yunnan) with palmately-veined stipulate leaves, the stipules being adnate to the slender petiole. The tracheids are of the same type as in *Trochodendron*; in the early wood they have very elongated scalariform bordered pits, which in the later wood give place to circular bordered pits (Fig. 4). The flowers are, however, of a specialised type, and in spite of the presence of a calyx are even more advanced than those of *Trochodendron*; they are small, tetramerous, syncarpous, and borne in slender catkins (Fig. 22); unlike those of *Trochodendron*, they are anemophilous.

The list of primitive flowering plants is not exhausted by the members of the *Magnoliales*, *Laurales* and *Trochodendrales*. We find primitive forms also in the orders *Piperales*, *Nymphaeales*, *Illiciales* (where the genus *Illicium* (Fig. 6, p. 55) is characterised by exceedingly primitive vessels), *Ranunculales*, and others. For our purposes, however, the families with which we have become acquainted in the preceding pages will suffice. We have reviewed a whole gallery of ancient angiosperms of varying degrees of primitiveness; and amongst them not a few are 'living fossils', archaic forms that have survived to the present day, e.g. some of the primitive *Winteraceae* and *Magnoliaceae*, the genus *Degeneria*, etc. Yet not one of these forms has remained at the ancient gymnospermous stage, and not one of them has retained the whole complex of primitive features; all of them have altered to some extent and become adapted to contemporary conditions. Indeed, in this way alone could they have survived.

CHAPTER 8

THE ORIGIN OF APETALOUS DICOTYLEDONS

In a special article devoted to the cycads, the famous German botanist Alexander Braun (1875) incidentally put forward (p. 374) some fundamental ideas which anticipated by several decades the basic tenets of phylogenetic angiosperm systematics. He pointed out the vistas that were opened in the systematics of the angiosperms if one applied to them the old simple morphological concept of the flower as simply a shoot, a cone or strobilus substantially homologous with the cones of the cycads. According to Braun, the primitiveness of the *Magnoliaceae* and related families is apparent when the flower is considered as a strobilus, and from this it follows that the '*Monochlamydeae*' (as they were called by De Candolle) or '*Apetalae*' (as they were termed by other authors), i.e. the families *Casuarinaceae*, *Betulaceae*, *Fagaceae*, *Myricaceae*, *Juglandaceae*, *Salicaceae*, *Urticaceae*, *Piperaceae*, *Chenpodiaceae*, etc.— which are characterised by small, simply constructed unisexual flowers—must be derivative, and that the simplicity of their flowers is not primitive but derived. To Braun belongs the well-known aphorism: 'In nature, as in art, the simplest may be the most perfected.' (See Kozo-Poljanski 1928, p. 45.)

Braun clearly understood that there are two kinds of simplicity—primary and secondary. However, in the science of his time, including his own system (1864) the *Apetalae*, because of the simplicity of their flowers, were regarded as the lowest group of dicotyledons. A competent scientific and historical evaluation of the concept of the '*Monochlamydeae*' as primitive was given by Kozo-Poljanski (1928) in his book *The Ancestors of the Flowering Plants*. 'The *Monochlamydeae*', he wrote (p. 44), 'were considered to be the lowest group of flowering plants because of the simplicity of their flowers and the affinity [in pre-evolutionary times "affinity" was understood to mean similarity in characters] with the conifers observed in their inflorescences. This was at a time when the idea of evolution had not yet become fully respectable in biology and the state of factual knowledge was very unsatisfactory, and when the need to distinguish the two types of simplicity of structure—primary simplicity, where a higher degree of organisation has never been attained, and simplicity by reduction, where a higher degree of organisation has already been attained and then lost—was still not realised. Moreover, no account was taken of the difference between the two types of *similarity*—essential similarity (homology) and superficial similarity (analogy).' Only a few biologists, like Darwin, understood fully and clearly that plants in which the various organs were most modified and reduced in number were more highly organised than those in which all the organs—sepals, petals, stamens and carpels—were fully developed in every flower. The difference between analogy and homology was far from universally understood. Unfortunately, the ideas of Alexander Braun, which he put forward only in passing, were overlooked by his contemporaries, and Braun himself was unable to develop them, as he died only two years later without having undertaken the reform of the system of classification of the angiosperms in accordance with the new principles. Later, but completely independently, Nägeli (1884) and Saporta (in Saporta et Marion, 1885) came to hold similar views, but the credit for reforming angiosperm systematics on an evolutionary basis belongs to Charles Bessey and Hans Hallier. Their first papers appeared in the 1890s, but the most complete account of Hallier's system appeared in 1912, and of Bessey's in 1915. Hallier was the leader of the new movement, and his system stands as the highest achievement of phylogenetic angiosperm taxonomy in the first quarter of the twentieth century. In 1916 Christopher Gobi's *Survey of the Systems of Plants* first saw the light of day in Petrograd; in it he published in full the system which he had already taught at the end of the 1890s in lectures to his students at St Petersburg University All three authors of these individually and independently constructed phylogenetic systems came to the conclusion that the '*Monochlamydeae*' were, in the first place, derivative and simplified and, in the second place, heterogeneous, the different members thereof having arisen along different evolutionary lines within the dicots.

We shall now consider a few groups of '*Monochlamydeae*' and attempt to demonstrate the derivative character of their organisation. We begin with the order *Piperales*, which in the Englerian system is considered to be one of the most primitive groups of dicotyledons. The absence of a perianth in the *Piperales* appears undoubtedly to be a secondary phenomenon, and the members of this order are not therefore primitively 'monochlamydeous', but had in the past a normally developed perianth. All the evidence points to the phylogenetic relationships of the *Piperales* being with the *Laurales* (especially with the *Chloranthaceae*) and also with the *Magnoliales* (especially the *Magnoliaceae*). We note, for example, that the majority of *Piperales* have stipules (like the *Chloranthaceae*), which in the genus *Piper* enclose the terminal bud, as in the *Magnoliaceae*. In the most primitive family in the order, *Saururaceae*, the vessels are very primitive, with scalariform perforation with numerous bars and scalariform or intermediate pitting in the side walls, features which bring them very close to the *Chloranthaceae* and *Magnoliales*. Even more significant is the presence in the *Saururaceae*, and in the *Piperaceae* of monocolpate pollen grains (Erdtmann, 1952), as in *Ascarina* and species of *Hedyosmum* in the *Chloranthaceae* and in the *Magnoliaceae*. The closeness of the *Piperales* to the *Magnoliales* is shown also by the apocarpous gynoecium of the primitive genus *Saururus*, in which, as serial sections of the flower have shown, the stamens and carpels are arranged in a spiral series (Raju, 1961), as is characteristic of the *Magnoliales*. The styles of the carpels of *Saururus* have a long decurrent stigma and both the ovary and the style are partly open (Murty, 1959), another sign of affinity with the magnoliaceous plexus. The order *Piperales* differs from the *Magnoliales* mainly in having perisperm in the seeds. Thus it is apparent that the *Piperales* are a group derived from the *Magnoliales* and have not yet lost their close connection with them.

The order *Caryophyllales*, also known as the '*Centrospermae*', is often included in the heterogenous group '*Monochlamydeae*'. This order has also been considered as primitive, and the possibility of its derivation from some group of '*Polycarpicae*' has been, and sometimes still is, denied; but Hallier, Gobi and many others derive it with good reason from the *Ranunculales*. The *Ranunculales*, including the *Lardizabalaceae*, *Menispermaceae*, *Ranunculaceae*, *Berberidaceae* and allied families, in their turn most probably arose (together with the allied order *Illiciales*) from the *Magnoliales*. Amongst the *Caryophyllales* the closest links with the *Ranunculales* are shown by the primitive family *Phytolaccaceae*, which has many features that bring it near the families *Menispermaceae* and *Lardizabalaceae*. We find within it forms with apocarpous gynoecia—such as *Anisomeria*, *Ercilla* and *Phytolacca* subg. *Pircunia*—which at once connect it with the *Ranunculales*. But the affinity with the *Ranunculales* is shown by the presence not only of apocarpous gynoecia, but also of tricolpate pollen grains of the ranunculalean type, with its characteristic granulate surface of the middle part of the aperture membrane. The presence of the curved embryo in the *Phytolaccaceae* and other *Caryophyllales* points to an affinity with the *Menispermaceae*. Stipules are sometimes found in the *Caryophyllales*, but they are also known in some *Menispermaceae*, and the secondary thickening from successive cambia (concentric type) of the stem of many *Caryophyllales* is also reminiscent of the *Menispermaceae*. Affinity with the *Ranunculales* is further corroborated by biochemical data—the presence of isoquinoline derivatives in the *Chenopodiaceae* and *Cactaceae*.

Also included by Wettstein in the artificial group '*Monochlamydeae*' is the order *Hamamelidales* (Figs. 23, 24), which is of especial interest, chiefly because it serves as a link between the *Magnoliales* on one hand and the '*Amentiferae*'—*Casuarinales*, *Urticales*, *Fagales*, *Myricales*, etc.—on the other. The *Hamamelidales* are very close to the *Trochodendrales*, *Cercidiphyllales* and *Eupteleales*, together with which they almost certainly arose from the *Magnoliales*. They have much in common with the *Magnoliales*, particularly in the structure of the secondary xylem (Tippo, 1938), but the flowers of most of them are rather specialised and characterised by well-marked simplification and reduction of parts, and by the frequent formation of capitate or catkin-like inflorescences; and this is associated with a gradual transition from entomophily to anemophily. In floral evolution the *Hamamelidales* have proceeded from

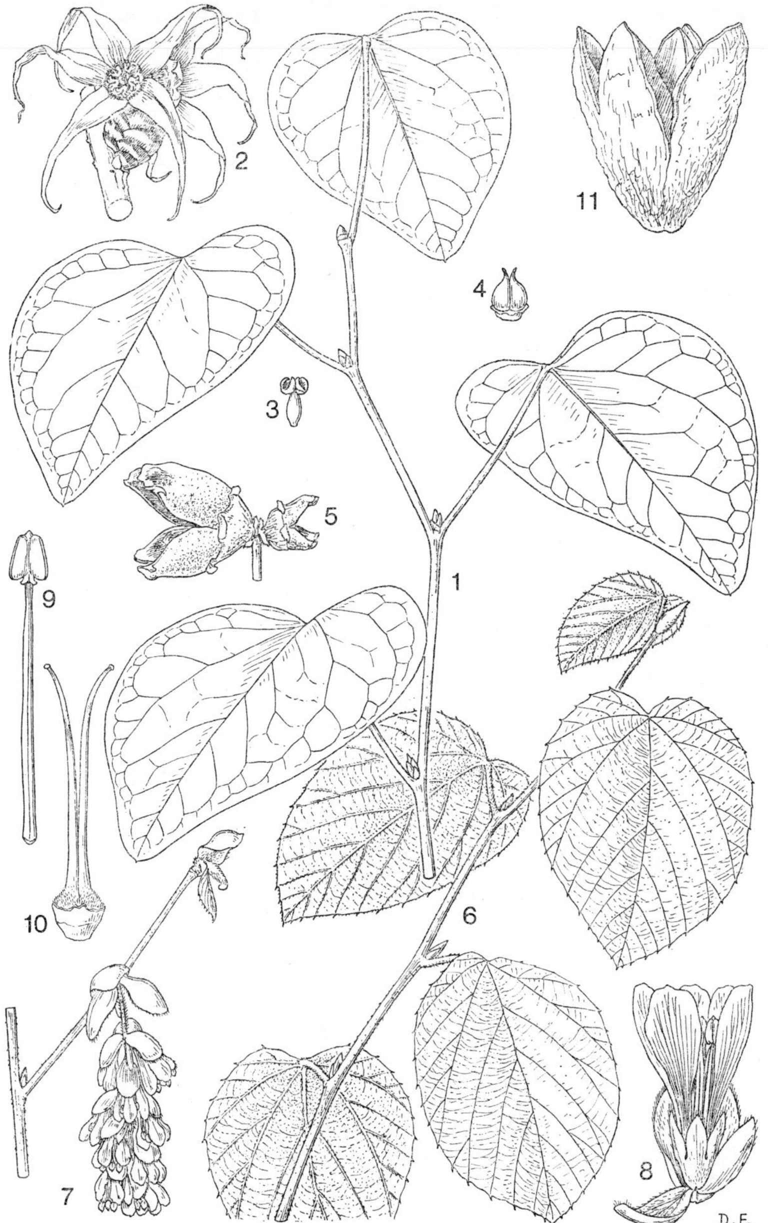

Fig. 23. *Hamamelidaceae*: *1–5*, a comparatively primitive member, *Disanthus cercidifolius*; *6–11*, a more advanced member, *Corylopsis spicata*. *1*, Leafy branch; *2*, pair of flowers; *3*, stamen; *4*, ovary; *5*, fruit, *6*, leafy branch; *7*, inflorescence; *8*, flower; *9*, stamen; *10*, ovary; *11*, fruit.

bisexual entomophilous types to unisexual apetalous anemophilous types, and from forms with apocarpous to forms with syncarpous gynoecia. Evolution of the pollen proceeded from tricolpate grains of the *Trochodendron* and *Tetracentron* type to pantoporate grains as in

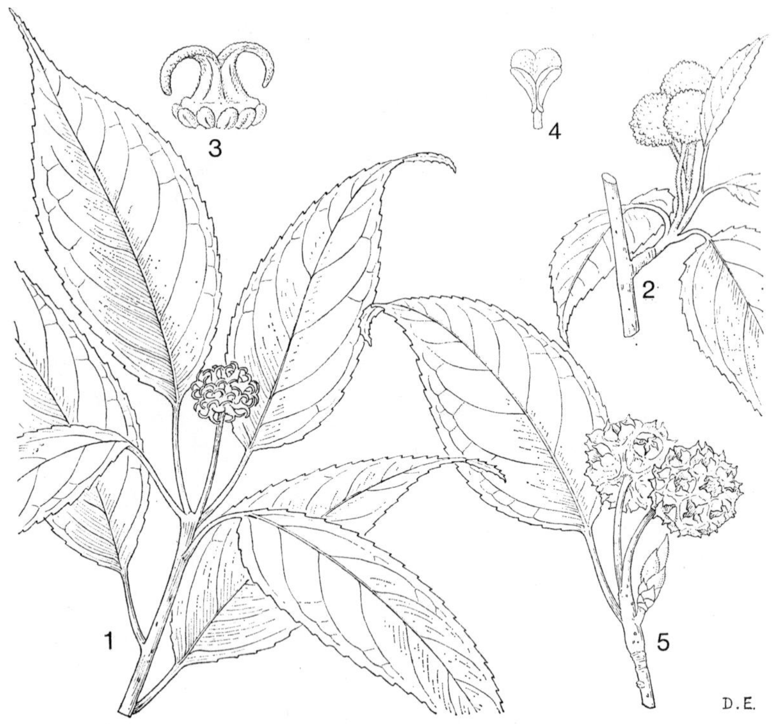

Fig. 24. *Hamamelidaceae*: an advanced member, *Altingia excelsa* (*Liquidambaroideae*). *1*, Female flowering branch; *2*, male flowering branch, showing heads of stamens; *3*, female flower; *4*, stamen; *5*, fruiting branch.

Altingia (Fig. 24) and *Liquidambar*. The primitive members of the *Hamamelidales*, and especially of the closely related order *Cercidiphyllales* (the genus *Cercidiphyllum*), are in many ways reminiscent of *Trochodendron* and *Tetracentron*. The haploid chromosome number in *Cercidiphyllum* is nineteen, i.e. the same as in *Trochodendron*, *Tetracentron* and *Magnoliaceae*, and the chromosomes in all these plants are short and rod-shaped. Solereder (1899) suggested that the *Hamamelidales* were linked with the *Trochodendrales* by the (in his opinion) intermediate genera *Cercidiphyllum* (Fig. 25) and *Eucommia* (Fig. 26/*1–7*, p. 103), and Hallier (1903) came to the

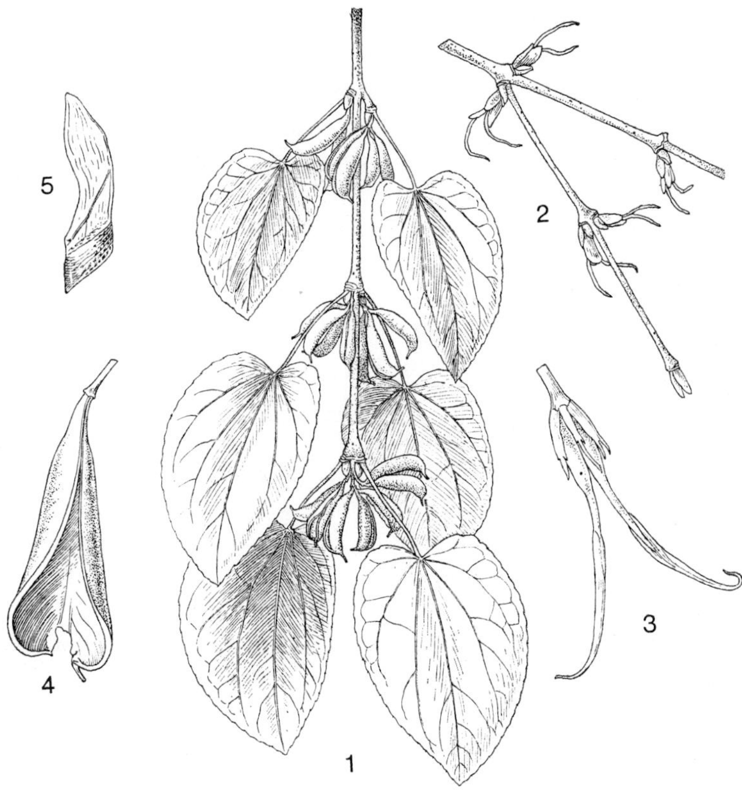

Fig. 25. *Cercidiphyllum japonicum*: *1*, fruiting branch; *2*, flowering branch, female; *3*, female flowers; *4*, fruit; *5*, seed.

conclusion that the *Hamamelidales* were a group intermediate between the *Magnoliales* in the broad sense and the *Amentiferae*. He referred the *Trochodendrales* to the *Hamamelidales*. Gobi (1916) in his system also derived the *Hamamelidales* (in which he included the *Hamamelidaceae*, *Cercidiphyllaceae*, *Eucommiaceae* and *Platanaceae*) from the *Magnoliales*. Kozo-Poljanski (1922) derived the *Hamameli-*

daceae from the *Tetracentraceae*. Croizat (1947) also emphasised the affinity of the *Hamamelidaceae* with *Tetracentron* and *Trochodendron*. They have much in common in gynoecial structure, particularly in the style with its primitive stigma more or less decurrent along the suture.

It also appears that the group of orders that forms the nucleus of the '*Monochlamydeae*' had its origin in the *Hamamelidales*; they are the ones often united under the name '*Amentiferae*'. It now seems to me possible to consider established with some certainty that they are a further development of wind-pollinated hamamelidaceous forms, in which the flowers have become even more simplified and the inflorescences still more adapted to anemophily. Closest to the *Hamamelidales* stand the *Eucommiales* (*Eucommiaceae*) and *Urticales*.

The family *Eucommiaceae* which consists of one monotypic genus *Eucommia* (Fig. 26/*1–7*) found in central and eastern China shows affinities, on one hand with the *Hamamelidaceae* and, on the other, with the *Ulmaceae*. Some, therefore, associate it with the former family, and others with the latter in the order *Urticales*; or at least they point out the possibility of such an arrangement. Solereder (1899) included *Eucommia* in the family *Hamamelidaceae* on the basis of the structure of the flower and wood, but at the same time showed it occupied a rather isolated position in the family, as the female flowers are solitary and the fruit structure quite anomalous. Moreover, its lacticiferous cells are unknown in the *Hamamelidaceae*, and in *Eucommia* scalariform perforation is found only in the primary xylem. *Eucommia* is closer to the urticalean families, especially the *Ulmaceae*. Oliver (1891) was the first to point out the possibility of an ulmaceous affinity because of a general similarity and a similar type of fruit; later writers also pointed out the similar type of leaf venation (Harms, 1930) and similarities in the structure of the wood, flowers, inflorescences, fruits and seeds (Tippo, 1940;

Fig. 26. *Eucommia* (*1–7*) and *Casuarina* (*8–12*), probably progressively more specialised unisexual derivatives of the *Hamamelidales*. *1*, Leaf and fruiting branch; *2*, fruit; *3*, female flower; *4*, *5*, male flowers; *6*, stamen; *7*, stamen in cross-section; *8*, male flowers; *9*, female flower-heads; *10*, female flower, ovary opened to show ovules; *11*, fruiting branch; *12*, fruiting head. (*8–12* After Poisson.)

12
9
10
11
8
1
2
3
7
4
5
6

Varossieau, 1942). Particularly important are, first, the marks of affinity in the anatomical structure of the secondary xylem, such as the simple perforation of the vessels, the usually alternate pitting of the side walls, the fibres with bordered pits and the level of specialisation of the rays, and, secondly, the occurrence of unicellular conical hairs with walls wholly or partly silicified or calcified. On the other hand, *Eucommia* is readily distinguished both from the *Hamamelidales* and the *Urticales* by a series of characters which include the absence of stipules, the unilacunar node, the tricolpate pollen grains somewhat similar to those of *Cercidiphyllum* (Erdtman, 1952), the single integument, the cellular endosperm (Eckardt, 1963), the presence of lacticiferous cells, and so on. Thus on the sum total of its characters the family *Eucommiaceae* constitutes a separate order (Němejc, 1956) which stands closest to the *Ulmaceae*, though at the same time in many respects it approaches the *Hamamelidales*. The *Ulmaceae*, the most primitive family of the *Urticales*, had in all probability a common origin with the *Eucommiales*.

The *Casuarinales*, consisting of the single family *Casuarinaceae* and one genus *Casuarina*, with the majority of its species in Australia, are also linked with the *Hamamelidales*. In the systems of Engler and Wettstein *Casuarina* is placed at the very beginning of the dicotyledons. According to Kuznetsov (1936), it is one of the most remarkable of the simpler types of angiosperm, standing quite isolated in the system and unconnected with any other family. But is this so?

At first glance, *Casuarina* (Fig. 26/*8–12*, p. 103) does indeed seem to stand quite apart from the other dicots, and the elucidation of its phylogenetic links with them appears extremely difficult. This apparent isolation of *Casuarina*, however, is a result of the peculiar and extreme specialisation of its leaves and reproductive organs, which causes it to look more like an *Ephedra* than an angiospermous plant. The unisexual anemophilous flowers, the entire absence of a perianth, the extreme reduction of the unistaminate male flowers, the tendency of the single stamen to split, the very specialised gynoecium, the reduction of the posterior locule of the ovary, the initiation of two to four ovules of which only one develops further, the fusion of the seed-coat with the pericarp, the absence of endosperm in the seeds, and the large embryo—all these are signs of high evolutionary development and far-reaching morphological specialisation. Specialisation is also apparent in the reduced scale-like leaves, in many anatomical features, and in the chalazogamy which is usual on the whole in dicots with simplified anemophilous flowers; the pollen—(2–)3(–5)–pororate—is also of a specialised type. But *Casuarina* is not so unusual and isolated as it appears at first sight; in many morphological features it approaches such families as the *Urticaceae*, *Betulaceae*, *Corylaceae*, *Myricaceae* and also the *Hamamelidaceae*. When Endlicher (1851) wrote of *Casuarina*, 'affinitate proximae sunt *Myriceis*' (p. 157), he was near the truth; others have placed the *Casuarinaceae* near to the *Urticaceae*, and with good reason. In many features of floral construction *Casuarina* undoubtedly comes near to the *Urticales*: the gynoecium consists of two median carpels and is pseudomonomerous, and the one-seeded samara recalls the samara of *Ulmus*; the two long stigmas in *Casuarina* are also characteristic of several members of the *Urticales*. On the other hand, in the structure and development of its gynoecium *Casuarina* approaches the *Betulaceae* and *Corylaceae* (Benson, Sanday and Berridge, 1905; Rendle, 1938, Eames, 1961). The pollen grains of *Casuarina* are very similar to those of the *Betulaceae* and *Myricaceae* (Erdtman, 1952, Kuprianova, 1965). In Hallier's system (1901, 1905, 1908, 1912a) *Casuarina* is not unreasonably placed between the *Corylaceae* and *Betulaceae* in the collective family '*Amentaceae*'. The similarity between *Casuarina* and *Betula* is explained, however, not by the origin of *Casuarina* from the *Betulaceae* but rather by their common origin from the *Hamamelidales*. Bessey (1915) derived the *Casuarinaceae* from the *Hamamelidaceae*, and this is accepted by Hutchinson (1926, 1959a) and supported by anatomical and morphological investigations (Tippo, 1938; Moseley, 1948). In wood anatomy *Casuarina* is more highly evolved than the *Hamamelidales*, and its flowers are a result of reduction and simplification of the hamamelidaceous type of flower. All the most important features of *Casuarina* can already be found in one or other member of the family *Hamamelidaceae*—unisexual anemophilous flowers, the structure of the inflorescence, reduction of the perianth, gynoecia of two median carpels, reduction in the number of ovules, and so on.

Investigation of the phylogenetic connections of the orders *Fagales*, *Betulales*, *Balanopales*, *Leitneriales* and other '*Amentiferae*' leads to a similar conclusion—that these orders also appear to be descendants of the *Hamamelidales*. All these orders are characterised by evident reduction in size and number of the flower parts and specialisation of the inflorescences associated with adaptation to wind pollination. Simplification of the individual flowers has been accompanied by increasing specialisation of the inflorescences, the structure of which has become considerably more complex than that of the ancestral type of flower (see, for example, Endress, 1967).

In conclusion, we consider the *Salicaceae* (willow family) which forms the order *Salicales*. The *Salicaceae* have always lacked a definite systematic position and have been put near such remotely allied families as the *Betulaceae*, *Fagaceae*, *Casuarinaceae*, *Juglandaceae*, *Myricaceae*, *Piperaceae* and *Hamamelidaceae*. Hutchinson (1926, 1959a) derived the order *Salicales*, like the '*Amentiferae*' from the *Hamamelidales*; but this is quite at variance with the findings of morphology. In gynoecial structure, and many other features, the *Salicaceae* are remote from the *Hamamelidales* and stand closest to the *Tamaricales* and especially the *Violales* (*Parietales*). Bartling (1830) placed the *Salicaceae* and *Tamaricaceae* together, and many later writers, including Bessey (1897), derived them from the *Tamaricaceae* or their immediate ancestors. In this the greatest emphasis has been placed on the presence of a hair-tuft (coma) on the seeds, the sub-basal position of the ovules in the carpel, and the identical dehiscence of the capsule. The similarity of the *Salicaceae* to the *Tamaricaceae* has apparently been somewhat over-estimated, however. They are most likely related, but they merely share a common origin from the pantropical and subtropical family *Flacourtiaceae*. Hallier (1908, 1911, 1912a) first suggested that the *Salicaceae* were most probably derived from the *Flacourtiaceae*, a conclusion accepted also by Gobi (1916) and Cronquist (1957). Studies of the wood anatomy showed them to be closest of all to the *Flacourtiaceae* (Gzyrian, 1952). Both in external morphology and in wood anatomy the greatest similarity to the *Salicaceae* is observed in the subtribe *Idesiinae* of the *Flacourtiaceae*, e.g. the monotypic Sino-Japanese genus *Idesia* (south Japan, Korea, continental China and

Taiwan). Study of the floral morphology and anatomy of the *Salicaceae* has made it clear that their ancestors must have had flowers of the flacourtiaceous type. Eichler (1878) and Velenovsky (1904) both suggested that the flowers of the *Salicaceae* arose from bisexual entomophilous flowers which had a normally developed perianth. This view is supported by the frequent occurrence of rudimentary organs in the flowers of *Salicaceae* and by the not infrequent instances of atavistic anomaly, in particular the development of bisexual flowers (Erlanson and Hermann, 1928).

Thus we are led to the conclusion that the so-called '*Amentiferae*' are a very heterogeneous and artificial assemblage, in which taxa of very different origins have been included. All these taxa are characterised by a reduction of the flower and inflorescence, associated with the change from entomophily to anemophily. Analogous processes of reduction can be observed in the other apetalous flowering plants.

CHAPTER 9

THE ORIGIN OF THE MONOCOTS

The division of the angiosperms into monocotyledons and dicotyledons, which dates from the time of John Ray (late seventeenth century), has persisted to the present day. The attempts made by a number of botanists, especially in the first quarter of the present century, to show that the monocots were not an evolutionary line but merely a grade of angiosperm development, have not been crowned with success.

The division of the monocots into several distinct unrelated groups which could be inserted at various places into the classification of the dicots has never been convincingly demonstrated, and therefore in the majority of contemporary systems of classification of the monocots are treated as a natural monophyletic developmental line. Their treatment as monophyletic is based on the undoubted kinship that exists between the component orders. They are a highly diversified group, and the extreme forms are very different from one another, but, just as such distinct families as *Lemnaceae* and *Araceae* in fact show their kinship through the genus *Pistia* (*Araceae*) which is in many ways intermediate, so all the monocots are linked together phylogenetically through various intermediate genera and form one branch of development having a single origin; see Hutchinson (1934, 1959b), Kimura (1956), Takhtajan (1959), Novák (1961) and Cronquist (1965).

Many different hypotheses have been put forward to explain the origin of the monocots, but at the present day it is practically universally accepted that they arose from the dicots. All the evidence indicates that they originated very early and that their ancestors were some of the most primitive dicots. Amongst the monocots there still exist forms with apocarpous gynoecia, such as *Butomaceae*, *Limnocharitaceae*, *Alismaceae*, *Scheuchzeriaceae*, *Potamogetonaceae*, *Triuridaceae* and some palms; and this is explicable if the monocots arose from some group of apocarpous dicots. The question is—what group?

The *Ranunculales* in the strict sense have most often been considered as possible ancestors of the monocots; they include such families as *Lardizabalaceae*, *Menispermaceae*, *Ranunculaceae* and *Berberidaceae* and others. Thus Hallier (1912a) derived the monocots from the *Lardizabalaceae*, while in Hutchinson's (1959b) opinion they arose from the *Ranunculaceae* (which he split into two families, *Ranunculaceae* and *Helleboraceae*). However, the pollen in *Ranunculaceae* is tricolpate, pantocolpate or pantoporate and is never monocolpate, although monocolpate pollen is especially characteristic of the majority of the monocots. Since the monocolpate is the most primitive angiosperm type, the monocots are consequently more primitive in their pollen than the *Lardizabalaceae*, *Ranunculaceae* and related families. The ancestors of the monocots must therefore have been dicotyledonous plants with primitive monocolpate pollen (Takhtajan, 1954a, 1961; Eames, 1961).

In the dicots, monocolpate pollen is known in the *Magnoliales* (Fig. 9, p. 60), some of the *Laurales*, the *Nymphaeales* (excluding *Nelumbo*), the majority of the *Piperales* and in the genus *Saruma* of the *Aristolochiaceae*, i.e. in comparatively few, and it is among these dicots, therefore, that we must seek the group which is closest to the extinct ancestors of the monocots. But it is necessary first to analyse the other characteristic features of monocotyledonous plants.

Comparative anatomical studies have shown that vessels originated independently in dicots and monocots (Cheadle, 1953). This is indicated by the fact that in the most primitive monocots vessels are either absent, as in the *Hydrocharitaceae*, or present only in the roots, as in *Butomaceae*, *Limnocharitaceae* and *Alismaceae*. In most of the remaining monocots, including the most specialised, there are vessels in other parts of the plant, often in all parts. Cheadle concluded from this that vessels first arose in monocots in the root metaxylem and later in the stems and leaves. It is in the roots that we find the most primitive types of vessels, having scalariform perforation with more than one hundred bars. Thus, if vessels in the monocots arose independently from those in the dicots, we must seek the ancestors of the monocots amongst the vesselless dicots (Takhtajan, 1954a, 1961; Eames, 1961); the families of the order *Ranunculales* amongst which there are no vesselless forms must therefore be excluded from the list of possible ancestors.

Amongst the apocarpous woody dicots, vesselless forms have persisted mainly, as we have seen, in the *Magnoliales*; and amongst herbaceous dicots only the *Nymphaeales* are most probably primitively vesselless. The *Nelumbonales*, consisting of the single genus *Nelumbo*, are characterised by tricolpate pollen and vessels and are therefore excluded from the list of possible ancestors. It is otherwise with the *Nymphaeales*, which have monocolpate pollen, and although its members lead an aquatic existence, the absence of vessels is probably not in this case a result of reduction. An aquatic habitat does not always lead to the loss of vessels; they are present, for example, in the roots of such typical aquatic plants as *Pontederiaceae* and *Potamogetonaceae*. Vessels might well have been retained in the *Nymphaeales* if they had been present in the past. But in the *Nymphaeales* vessels are completely absent, and even in the substantial rhizomes of the *Nymphaeaceae* we find in the metaxylem only long scalariform tracheids of a wholly primitive type. There is therefore every reason to accept the order *Nymphaeales* as a primitively vesselless group. If this is so, then in the search for the group closest to the actual ancestors of monocotyledons we have only two orders to choose from, the *Magnoliales* and the *Nymphaeales*.

Amongst the monocots, true woody plants—trees and shrubs such as we find in the dicots—are completely absent. Palms, *Pandani*, some *Agavaceae*, bamboos and other arborescent monocots are neither trees nor shrubs but are special growth forms that have arisen in the course of evolution only amongst the monocotyledons. Studies in comparative anatomy have led many writers to conclude that the arborescent monocotyledonous forms have been derived from herbaceous ancestors. Thus, according to Eames (1961), the unbranched arborescent palms have arisen from freely-branching herbaceous rhizomatous palms, and he suggests an analogous course of evolution for the bamboos. It must be emphasised, however, that although palms and bamboos may be arborescent plants they are not truly woody plants. As Alexandrov (1954, p. 227) pointed out, the difference between woody and herbaceous stems lies not in consistency but in structure. He writes, 'Thus the species of palms, which sometimes have extremely hard tissues forming the bulk of their stems, are in fact structurally herbaceous plants.' In those cases, such as *Agave*, *Dracaena* and *Yucca*, where secondary thickening of the stem does occur, it is brought about quite otherwise than in the dicots (Cheadle, 1937; Alexandrov, 1954, Esau, 1965): a meristematic ring resembling a cambium arises outside the vascular bundles. As Tomlinson (1964) says, 'The cambium which adds secondary tissue to the primary axis is not an initial cambium, as in a dicotyledon, but a tiered cambium' (p. 76). The primitive monocots of the order *Alismales*, like the most primitive members of the central order *Liliales*, are typically herbaceous forms, in many respects very close to primitive members of the *Nymphaeales*. From this we may conclude that the immediate ancestors of the monocots were most likely some long extinct vesselless herbaceous plants with apocarpous gynoecia and monocolpate pollen which probably had much in common with the modern *Nymphaeales* (Takhtajan, 1943, 1959, 1961).*

As long ago as 1905, Hallier (1905, p. 157) suggested that the *Nymphaeaceae* (*s.l.*) were the 'ancestors of *Helobiae* and of the whole division of monocotyledons', though later (Hallier, 1912a) he changed his opinion. According to Agnes Arber (1920, p. 309), the *Nymphaeaceae* 'descended from a stock closely related to that which gave rise to the monocotyledons'. And Parkin (1923, p. 58) says, 'The floral features in common between the *Helobiae* on the one hand and the *Ranales* (especially certain of the *Nymphaeaceae*) on the other hand suggest something deeper than the mere parallelism.' Analogous ideas have also been expressed by some other botanists.

The primitive monocots and the order *Nymphaeales* do indeed have much in common. In the structure of their apocarpous gynoecia the families *Butomaceae* and *Limnocharitaceae* resemble the *Cabombaceae*, and in their 'diffuse' placentation they recall the

* In the first version of my diagram of the phylogenetic interrelationships of the flowering plants, published in 1942, I derived the monocots directly from the *Nymphaeales*, but in later versions I have preferred to derive them from their immediate ancestors.

Nymphaeaceae. Also, in their stems with their scattered closed vascular bundles (atactostelic vascular cylinder), in the more or less reduced primary root and in root structure, the *Cabombaceae* and *Nymphaeaceae* have much in common with the monocots.* There are also some features of leaf structure and development common to the *Nymphaeales* and *Alismales*. One may instance the interesting similarity in early leaf development between certain of the *Nymphaeaceae* (e.g. *Victoria*) and the *Alismaceae*. According to Nitzschke (1914), *Cabomba* and *Brasenia* show the greatest similarity, so far as the development of the female gametophyte is concerned, to the *Butomaceae* and *Alismaceae*, and according to Earle (1938), the *Nymphaeaceae* are very close in their embryology to typical monocots and are clearly different from the *Magnoliaceae*, *Ranunculaceae* and *Berberidaceae*. Their sporoderm and stomatal patterns are also monocotyledonoid (Snigirevskaya 1964; Meyer 1964, 1966; Dunn, Sharma and Campbell, 1965) as well as the arrangement of the first leaves (prophylls) on lateral axes (Fries, 1911).

It is not surprising, therefore, that various authors have regarded the *Nymphaeaceae* as monocots (Trécul, 1845; Seidel, 1869; Schaffner, 1904 and Cook, 1906) and Schaffner (1929, 1934) found it possible to transfer the order *Nymphaeales* to the monocots, where he placed it between the *Alismales* and *Triuridales*. The only reason preventing this taxonomic transfer is the presence in the *Nymphaeales* of two cotyledons. But whether the *Nymphaeales* are put in the monocots or in the dicots, their affinity with the *Alismales* cannot be doubted. Their close connection with the primitive dicots is no less certain; there is every reason for considering the order *Nymphaeales* as a hydrophilous derivative of some ancient *Magnoliales*, the vegetative organs of which have more or less degenerated in an aquatic environment. Such ancient herbaceous dicots of nymphaealean type, but more primitive and less reduced

* Whereas in dicots the protoderm arises as the innermost layer of the tunica, and thus has a common origin with it (the climacorhizous type of Van Tieghem), in the monocots (and in the *Nymphaeales*) it arises as the outermost layer of the corpus (the liorhizous type of Van Tieghem)—see Van Tieghem and Constantin (1918) and Voronin (1964) for details. Furthermore, it has been shown that the liorhizous roots of the *Nymphaeales* are not of an intermediate type; they have the characteristic structure of monocot roots (Voronin, 1964, p. 65).

than their present-day representatives, quite probably initiated the monocotyledonous line.

In comparison with the majority of dicots, the monocots, including all primitive taxa, are characterised by some degree of 'infantilism' of their vegetative organs. They usually have no vascular cambium or some of the bundles have a weak and short-lived cambium, their primary root is usually ephemeral and is replaced by a number of adventitious roots, the leaves are not differentiated into petiole and blade or the differentiation is not as clear-cut as in the dicots, the leaf venation resembles the venation of the under-developed phyllomes of the dicots (like bud scales or bracts), and one of two cotyledons is suppressed. On this basis I suggested that neoteny was involved in the origin of the monocots (Takhtajan, 1943, 1954a, 1954b). But as is well known, neoteny is usually connected with extreme environmental conditions, such as a surplus or deficiency of water, low temperature, etc. What kind of environmental conditions brought about neoteny in the ancestors of the monocots?

It has long been suggested that the monocots arose as a result of a hydrophilous or hygrophilous evolutionary tendency in their dicot ancestors. Henslow (1911) considered the monocots arose from aquatic dicots, while Jeffrey (1917, p. 415) suggested an aquatic or amphibious way of life might have led to the loss of cambial activity, just as in the *Nymphaeaceae* the scattered arrangement of vascular bundles and the absence of cambial activity are correlated with an aquatic existence. Bews (1927, p. 53) wrote that the earliest monocotyledons were either marsh or forest-margin types. It is a fact, as a survey of aquatic and marsh dicots shows, that under the influence of an aquatic environment many dicotyledonous herbs have developed distinct monocotyledonous features in their vegetative organs. The overall result of the action of an aquatic environment on the structure of the plant is a general degeneration involving marked simplification of the vegetative parts. Many parts and tissues become more or less arrested in their development, adaptations to life on land, unnecessary in the water, disappear, the development of the main root is suppressed, the anatomical structure of all vegetative parts is simplified and their conducting systems reduced, air chambers are formed, and so on. Henslow's hypothesis of the aquatic origin of monocots has been criticised by Ethel Sargant (1903, 1904 and 1908), who concluded in her turn that many of the characteristic features of the monocots may be easier explained as having arisen as a result of adaptation to a geophilous habit. But apparently nearer to the truth was Parkin (1923, p. 59), who suggested 'the golden mean' between the two hypotheses. He writes: 'Respecting the relative merits of an aquatic or geophilous ancestry for monocotyledons, the two views may be somewhat reconciled by regarding the earliest ones as neither markedly aquatic nor extremely geophilous—in fact, marsh plants with stout rhizomes. Some of their descendants have become completely hydrophytic, others sharply geophytic, while others again have retaken to the arborescent habit by fresh means.' Indeed, the majority of *Alismales*, as well as almost all the *Nymphaeales* (except *Ceratophyllum*), are typical 'aquatic geophytes' as Arber (1920, p. 217) calls them. Apparently the common ancestors of both the *Nymphaeales* and monocots were amphibious geophytes in which geophily at first arose under terrestrial conditions—most probably under the forest canopy or in the forest margin.

One of the most characteristic features of the monocots is the embryo, which has only one cotyledon. While the great majority of dicots have two cotyledons, in the monocots only one develops. We must therefore consider the origin of the monocotyledonous embryo.

Hegelmaier (1874, 1878), on the basis of the comparative study of various dicot and monocot embryos, suggested that the monocot embryo arose as a result of the failure of one of the two cotyledons of the typical dicot embryo to develop. This, the so-called abortion hypothesis (*Abort-Hypothese*), was subsequently widely propagated, being developed by Henslow (1893, 1911) Winkler (1931), Metcalfe (1936), Yakovlev (1946), Eames (1961), Solntzeva and Yakovlev (1964) and many others. However, it also had many opponents.

Johansen (1945) strongly criticised the abortion hypothesis; in his opinion, the two cotyledonary growing points found in the majority of dicots are arranged symmetrically in relation to the embryonic axis and do not have their analogue amongst the monocots at any stage in their embryonic development. He regards the growing point of the single monocot cotyledon as situated on the embryonic axis itself and maintains that there is no other corresponding growing point. He concludes from this that the absence of a second growing point can be explained simply by the fact that no such growing point ever existed. Johansen considers therefore that the abortion hypothesis cannot on principle be expected to explain the origin of the cotyledon of monocotyledonous plants.

The interpretation of the monocot cotyledon as a morphologically terminal structure has received wide currency in the literature, being supported in particular by Souèges (1954) and shared by the present author in his previous works (till 1961). Doubt as to the morphologically terminal nature of the monocot cotyledon has long been expressed, however; Hegelmaier considered its terminal position to be a result of displacement, and many later writers, amongst them Kozo-Poljanski (1922, p. 128), also considered that the apparently terminal position of the cotyledon might be explicable in terms of displacement during growth. Many special investigations, the results of which have been published over the last fifteen years, especially those of Haccius (1952, 1954) and Baude (1956), have provided convincing evidence to show that the apparent terminal position of the monocot cotyledon is a result of displacement of the apical meristem by the strongly-growing cotyledon; the extent to which the cotyledon assumes a terminal position was found to vary in different monocots. This conclusion is also supported by Meyer (1960), Esau (1960, 1965) Eames (1961) and others.

We may conclude, then, that the monocot embryo arose as a result of the suppression or loss of one of the two cotyledons and we are thus back, though on a different level, to the old abortion hypothesis of Hegelmaier. Confirmation of it is found first of all among the dicots, where several species, or even whole genera, normally have monocotyledonous embryos; they include *Ranunculus ficaria*, some species of *Corydalis*, *Bunium*, *Peperomia*, *Cyclamen*, *Claytonia*, *Pinguicula*, some *Gesneriaceae* of the tribe *Cyrtandrae*, etc. (Hegelmaier, 1878; Sargent, 1903; Hill, 1906, 1907, 1920, 1938;

Metcalfe, 1936; Crété, 1956). The second cotyledon may be present as a rudiment and capable of development (*Cyclamen*) or may be completely suppressed, as in *Ranunculus ficaria*.

Studies of the early stages of development of many monocot embryos have also supported the origin of monocotyledony by reduction and loss of one cotyledon. Many investigations, including the most recent, have clearly shown that the earliest stage of development, the formation of a two-celled proembryo, is the same in dicots and monocots. This initial proembryo stage is the most conservative in embryo development and, as a result, is much the same even in the phylogenetically remotest groups of angiosperms. Further development proceeds in different ways; in dicots a lateral cotyledon arises on each side of the stem primordium, while in the monocots only one cotyledon develops which is displaced towards the apex, the stem primordium arising at the side. But initially in the monocots the primordia of two cotyledons are more or less clearly demarcated, though one is arrested in development and becomes indistinguishable in later stages. Records of numerous researches, scattered in the literature, have clearly shown how the monocot embryo becomes asymmetrical and the single cotyledon pseudoterminal. In *Hosta glauca* (*Liliaceae*), studied by Hanstein (1870), the typical spherical stage with its characteristic radial symmetry is followed by a stage in which furrows are formed in the apical zone of the proembryo and two projecting primordial swellings are initiated. At this stage the proembryo of *Hosta* differs little from that of dicots such as *Capsella bursa-pastoris*. In the later stages of development, however, *Hosta* diverges sharply from the dicot type. The symmetrical vertically growing proembryo becomes asymmetrical, and in consequence the primordial swellings of the future cotyledons are shifted to one side. This shift in plane of the cotyledon primordia results from a distinct inequality of growth in the cells on two opposite sides of the proembryo. This inequality continues in later growth, so that the furrows between the primordia of the two cotyledons, originally situated at the apex of the proembryo, come to occupy the middle of the side on which the growth of the cells is retarded. As a result, the apical zone of the proembryo with its two cotyledon primordia takes up a lateral position. This shift of the cotyledonary swellings is soon completed, but the pseudoterminal cotyledon begins to grow strongly, while the primordium of the second cotyledon ceases to grow (see also Yakovlev, 1946). The whole developmental process of the embryo of *Hosta glauca* thus clearly indicates that the monocot embryo arose from the dicot by way of a distinct alteration in the early stages of development.

No less interesting is some recent work on the embryogenesis of two members of the family *Hydrocharitaceae*—*Ottelia alismoides* (Haccius, 1952) and *Stratiotes aloides* (Baude, 1956). In its early developmental stages the embryo of *Ottelia* is characterised by two lobes which are the cotyledon primordia; one of them rapidly outgrows the other and comes to occupy the pseudoterminal position, while the other markedly lags behind and becomes placed below it. From this second lobe the first leaf eventually arises, while the second leaf is formed between it and the apical cotyledon, and it is here also that the growing point of the stem is situated (Haccius 1952, p. 450-452). A reduced second cotyledon is found in some monocots, as, for example, in the *Dioscoreaceae* (Beccari, 1870; Lawton and Lawton, 1967). According to Lawton and Lawton (p. 159), all five species of *Dioscorea* investigated by them have two cotyledons, one of which remains in the seed as an absorptive organ and one of which, the emergent cotyledon, develops later, appears above ground and carries out the normal photosynthetic functions of a leaf.

The second characteristic feature of the monocots is the almost complete reduction of the vascular cambium. However, in many monocots a weakly developed cambium can be seen in the free vascular bundles (Arber, 1925; Eames, 1961) and this is rightly regarded as an ancestral trait indicating that the ancestors of the monocots possessed an active cambium. According to Sargant (1903, p. 81), the distribution of vascular bundles in the stem and the loss of cambium in the monocots may be ascribed to the shortening of internodes and thickening of the subterranean axis. The reduction of the cambium, the parenchymatisation of the shortened and thickened stem, and the great development of the leaf-bases have led to the characteristic monocotyledonous arrangement of the vascular bundles, which in a cross-section of the stem are scattered over all or almost all the surface, or arranged in two—or several—more or less concentric circles, and only rarely form a single ring. These peculiarities of monocot stems distinguish them from the great majority of dicots; but in the *Nymphaeaceae*, many *Ranunculaceae* (*Actaea*, *Thalictrum*, and some species of *Anemone*), *Podophyllum*, *Leontice*, *Bongardia*, *Peperomia* and some other herbaceous dicots the vascular bundles have a more or less reduced cambium and are arranged in the monocotyledonous manner.

In the structure of their stems the monocots are in many respects a step ahead of the herbaceous dicots. 'The monocotyledons in fact represent the herbaceous type in its extremest form,' wrote Jeffrey (1917, p. 415). In the monocots the parenchymatisation of the stem has reached its maximum and the vascular bundles are always discrete, running from node to node without coalescing. The axial conducting system of the monocots achieves maximum discreteness and, freed from the rigid framework of the vascular cylinder, the numerous leaf traces can be distributed over the entire cross-section of the stem in the most expedient manner. This change in the arrangement of the vascular bundles is made possible only through the loss of the ability to produce secondary growth.

The leaves of most monocots have a venation which is more or less closed at the leaf apex; the veins which emerge from the leaf base usually run together again at their apices. The secondary (lateral) veins are usually much more weakly developed than the primary (longitudinal) veins and are often more or less reduced. In many palms, aroids, gingers and some others the venation again becomes open in the upper part of the leaf, but, these cases apart, unlike that of the majority of dicots, the entire system is closed, and blind endings of the lowest order veinlets in the areolae are not observed.

The leaves of the monocots in all probability arose from simple entire leaves with pinnate venation that were clearly differentiated into petiole and blade. The intermediate type of leaf probably had arcuate-pinnate venation and a broad sheathing base. It gave rise to the leaf with arcuate-parallel venation from which there arose in turn the leaf with the so-called 'parallel' venation. Arcuate-parallel, and particularly parallel, venation are characteristic mostly of monocotyledonous leaves, although they are also found (especially the arcuate) in some dicots. The peculiarities of monocot leaves are evidently associated with the way of life of their ancestors, which were probably rhizomatous herbs of marshy places. According to Sargent (1903), the shape and venation of monocot leaves are the most suitable for attachment to a shortened axis and for penetrating through the soil. Henslow (1893, 1911), on the other hand, explained the peculiarities of the leaves of monocots as being due to the influence of an aquatic environment.

In the monocots, the primary root soon ceases to function and is replaced by adventitious roots. True, there are monocots in which a weak development of the main root is observed, and in many cases the primary root of the embryo develops comparatively strongly, but this only shows that the characteristic fibrous root system of the monocots is a secondary evolutionary development. It is interesting also that in many dicots, e.g. *Ranunculus ficaria*, the main root does not develop, and this lack of development is well-known to be associated in many cases with hydrophily or geophily. According to Sargant (1903, p. 82), the suppression of the primary root and the development of the annual crop of adventitious roots are clearly bound up with the annual recurrence of a period of vegetative activity in geophilous plants.

The majority of monocots have trimerous flowers, thus differing from the great majority of dicots which are usually pentamerous or tetramerous, but resembling on the other hand the flowers of many primitive dicots. There are no essential differences between the flowers of the monocot order *Alismales* and the flowers of the family *Cabombaceae* of the order *Nymphaeales* (Fig. 27).

The evolution of the monocots began with forms having an apocarpous gynoecium of primitive carpels containing many ovules. We find the most primitive type of flower in members of the *Butomaceae* (Fig. 27) and *Limnocharitaceae*. These two families together with the family *Alismaceae* make up the order *Alismales*, which is one of the most primitive amongst the monocots. Nevertheless, the order *Alismales* also has many specialised features, such as the absence of endosperm in the seeds and the pantoporate pollen of most of its members, and should therefore be regarded as an ancient side-branch of monocot development and not as a basic ancestral group—see Takhtajan (1959) and Cronquist (1965). Primitive

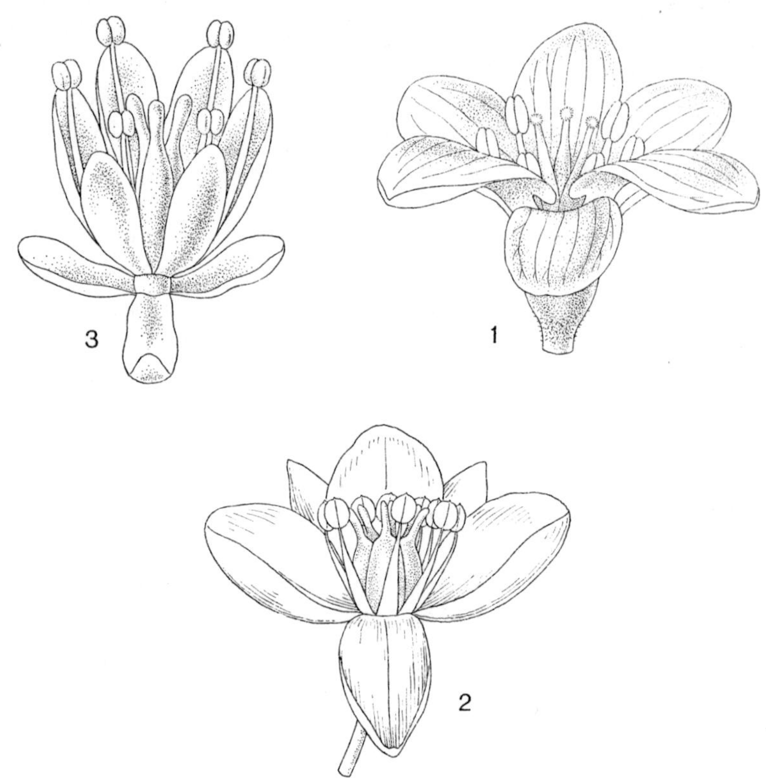

Fig. 27. *1*, Monocotyledonoid flower of the dicot *Cabomba aquatica* (*Nymphaeales*). *2*, *3*, Flowers of primitive members of the monocotyledons; *2*, *Butomus umbellatus* (*Alismales*); *3*, *Tofieldia calyculata* (*Liliales*); note the similarity of plan. (*1* After J. D. Hooker; *3* after W. J. Hooker.)

features are found also in members of several other monocot orders. The *Triuridales* and the genera *Tofieldia* (Fig. 27) and *Veratrum* (*Liliaceae* subfam. *Melanthioideae*) are also characterised by the primitive structure of their carpels and gynoecium (El-Hamidi, 1952; Eames, 1961), *Tofieldia glutinosa* having an almost apocarpous gynoecium. The gynoecium of the *Melanthioideae* is in many ways reminiscent of the gynoecium of the *Alismales* (and of *Scheuchzeria* of the *Najadales*), though on the whole it is more specialised. On the other hand, the *Liliales* are characterised by the presence of abundant endosperm and a small embryo. Primitive features (including apocarpous gynoecia) are also present in some palms. No single order amongst the monocots, however, combines in itself all the most important primitive features of this line of angiosperm development. In other words, there is no monocot order that could occupy the place in their family tree that is occupied by the *Magnoliales* among the dicots. A possible explanation of this is a more rapid rate of evolution in the monocots than in the lower, woody, dicots.

CHAPTER 10

FLOWERING PLANTS—THE FIRST TRACES

The study of even the most primitive living dicots cannot tell us when the angiosperms first made their appearance. This question could be answered by palaeobotany, if the very earliest links in the chain of angiosperm evolution were in fact known to science; but in spite of the striking successes of palaeobotany over the past few decades the earliest angiosperms remain as much a mystery to us as they were in Darwin's time.

The lack of palaeobotanical knowledge of the earliest angiosperms may probably be the result of their having been for a long period of time limited in their distribution and having long played a very insignificant role in the vegetation of the world. Moreover, as Seward (1933, p. 322) wrote, 'It is probable that the almost complete absence of fossil angiospermous leaves in Jurassic and older Mesozoic rocks is due, not to the lack of flowering plants in the world, but to their failure to be preserved as fossils because they occupied a tract of country remote from localities where the conditions were favourable for fossilisation.' If the hypothesis of their montane origin is correct, then they did in fact grow for some time in conditions extremely unfavourable for burial and fossilisation. Whatever the reason, the prehistory of the angiosperms remains unknown and the chances of filling this annoying gap in our knowledge are indeed small.

The angiosperms undoubtedly originated long before the Cretaceous period. The specialised character and astonishingly modern facies of many Cretaceous angiosperms 'confirm our belief in an antiquity of angiosperms antedating by many millions of years, probably by several geological periods, the first appearance of recognisable pioneers of the present ruling dynasty in the modern world' (Seward, 1933, p. 400). Arldt (1907, p. 581, 1938, p. 943) suggested that the angiosperms arose in the Trias, and Berry (1920) likewise suggested this period as the time of their first appearance, as can be seen from his proposed geochronological plan of the lines of descent of the higher plants. Several later writers postulated a Triassic or even Permo-Triassic origin for the angiosperms (Wieland, 1933; Camp, 1947; Thomas, 1947; Just, 1948; Axelrod, 1952, 1960, 1961; Zimmermann, 1959; Hawkes and Smith, 1965), while some others, e.g. Scott (1911, 1924), Golenkin (1927) and Němejc (1956), even referred their origin to the Permo-Carboniferous. Just (1948, p. 101) was quite correct in considering that 'in the light of known rates of evolutionary processes, a much greater age must now be postulated for the earliest angiosperms'. Eames (1961, p. 469) comes to the same conclusion. An Early Mesozoic origin seems the most likely for the angiosperms; they could scarcely have attained such a high morphological diversity if they had arisen later than the Trias. As yet, however, we have no direct evidence.

The angiosperm remains that have been described from pre-Cretaceous deposits are very problematical and have given rise to much controversy—see particularly Kräusel (1956), Scott, Barghoorn and Leopold (1960), Axelrod (1961) and Andrews (1961). Pollen finds have been described most frequently, especially from Jurassic deposits. Thus, they have been reported by various writers from the Jurassics of Scotland, Sweden, Germany and the U.S.S.R.; but they have been isolated finds of very few pollen grains; and, what is more, their angiospermous nature is far from indisputable. Pollen grains from the Middle Jurassic of Scotland, for example, considered to represent angiosperms of the *Nymphaea*, *Nelumbo* and *Magnolia* type (Simpson, 1937), proved to be pollen grains of gymnosperms (Hughes and Couper, 1958). Likewise, pollen grains from the Lower Jurassic of Sweden, described under the name *Eucommidites*, turned out to be gymnospermous (Couper, 1958). Other reports of the finding of angiosperm pollen grains in pre-Cretaceous deposits are likewise erroneous, or at least dubious, especially such reports from deposits of Palaeozoic age. The rare isolated leaf impressions from Triassic and Jurassic deposits which have been attributed to angiosperms are no less disputable. For example, Lignier (1908) described a leaf impression which he called *Propalmophyllum liasinum*; but this impression, which has a superficial resemblance to the leaves of the fan palms, is most likely a gymnosperm of the cycad or bennettitalean type (Scott, Barghoorn and Leopold, 1960, p. 286). The similarity between the fan palms and the leaf impressions described from the Triassic of Colorado under the name *Sanmiguelia lewisii* (Brown, 1965) is likewise superficial. Some writers, such as Andrews (1961), consider *Sanmiguelia* as almost certainly an angiosperm; others support its reference to the gymnosperms (Scott, Barghoorn and Leopold, 1960, p. 287), while to me it seems more likely that it is unrelated to the seed plants. Samylina has suggested (oral communication) that the leaf-remains, described under the name *Sanmiguelia*, probably belong to a gigantic species of *Schizoneura* (*Equisetales*); species of this genus with large leaves are known already (e.g. *S. ferganensis* Krysht.). There is also an impression of an entire pinnately-veined leaf of the dicot type from the Lower Jurassic of Germany (Kuhn, 1955), which is, however, unique and only poorly preserved; without further finds, it is thus impossible to pronounce with certainty on its systematic position. The possibility that it belonged to some extinct gymnospermous plant cannot be excluded. Thus, indisputably angiospermous remains of pre-Cretaceous age are not yet known to science; the situation would be quickly altered if we succeeded in finding together several different leaves of angiosperm type or even one characteristic angiosperm fruit.

Fully authentic angiosperm leaf remains (Plate XIII) and fruits are found only from the Early Cretaceous onwards. From the Neocomian (beginning of the Early Cretaceous) we know in fact only a few absolutely authentic angiospermous remains. Indisputable dicot fruits have been discovered in the Valanginian deposits of Vaucluse in southern France (Chandler, 1958), and a similar discovery of slightly younger (Hauterivian) fruits (*Onoana californica*) has been reported from Northern California (Chandler and Axelrod, 1961). Some indisputably angiospermous leaf remains are also known from some other Early Neocomian localities in North America.

From the second half of the Early Cretaceous (Aptian and Albian) considerably more certain remains of angiosperm leaves are already known, and there are also remains of dicotyledonous wood, although such finds are still comparatively rare. More than fifty years ago fossil angiosperm woods of rather unusual types were found in the Upper Aptain (Lower Greensand) deposits of southern England.* One of them (*Woburnia*) is thought to belong to the palaeotropical family *Dipterocarpaceae*, and another (*Aptiana*) is evidently a representative of the family *Theaceae*. The systematic positions of the other woods are still uncertain; but as Stopes (1912, 1915), who first described them, pointed out, not one of them is primitive in structure; they are all woods of a specialised type.

Though the dating of these southern English woods remains doubtful, the age of the angiosperm leaf remains discovered in the last century in the reputedly Upper Aptian deposits of Cercal, near Torres Vedras in Portugal is more certain. Of the eighteen species of this fossil flora the Portuguese palaeobotanist Teixeira (1948, 1950) refers four to the angiosperms. Not one of these four species can be referred to any existing genus and they are placed, therefore, in so-called 'form-genera' or 'organ-genera'.

Within the U.S.S.R., Aptian angiosperms are known from the Nikanskii deposits in southern Primorie, in the Suifun basin, where the monocot *Pandanophyllum ahnertii* was discovered, and in the North Suchan suite from which the dicots *Aralia lucifera*† and the angiospermous fruits *Onoana nicanica* and *Nyssidium orientale* (Plate XIII/5) have been described. These few angiosperms are accompanied by a purely Early Cretaceous flora consisting of ferns, cycads, bennettitaleans and conifers,—about one hundred species in all (Kryshtofovich and Prynada, 1932; Kryshtofovich, 1957; Krassilov, 1965, 1967; Samylina, 1960, 1968).

Towards the end of the Early Cretaceous the frequency and abundance of angiosperms noticeably increase. Angiosperm leaves

* However, angiosperm pollen is quite absent from the Aptian deposits of Great Britain (Couper, 1958), and this makes one doubt the correctness of the dating of these woods.

† *Aralia lucifera* Krysht. has lobed leaves and cannot therefore be referred to the genus *Aralia*, all members of which are characterised by pinnate leaves.

have been found on a fair number of occasions in deposits of Albian age in various parts of the northern hemisphere, including the Arctic, where Aptian angiosperms are still unknown. In the Albian fossil floras angiosperms still appear in the company of cycads, bennettitaleans, ginkgos, conifers and ferns, and predominate only in a few cases, as in the floras of Nazaré in Portugal and Dakota and Aspen in the U.S.A. The earliest of these mixed Albian floras in Europe are the floras of Buarcos and Tavarede in Portugal. In the former of these, angiosperms make up about one third of the species, but in the latter nearly half. The angiosperms are referred in most cases to form-genera, except for *Magnolia delgadoi* and *Aralia calomorpha*.*

A few Albian fossil floras are known from the U.S.S.R., the earliest being evidently the extremely interesting floras of the middle Zyrianka (a left-bank tributary of the River Kolyma in north-east Siberia) and Omsukchan (a village on the right bank of the River Kolyma), studied by Samylina (1959, 1960, 1968). Angiosperms occupy a subordinate position in the flora, being found only rarely, and in most cases as solitary impressions against a background of an overwhelming majority of fern and gymnosperm leaves. In number of species they total about 25%, but in number of impressions (and therefore probably in number of individuals) they form an insignificantly small proportion of the total fossil flora. The predominant plants are still ferns, cycads, ginkgos and conifers. In all, Samylina found twenty-two species of angiosperms, the majority of which she referred to form-genera or organ-genera. Some of these species are, however, of extreme interest, as they are almost certainly referable to such purely herbaceous families as *Ranunculaceae* (*Ranunculicarpus*, Plate XIII/3) and *Cyperaceae* (*Caricopsis*), thus once more confirming the great antiquity of herbs in both monocots and dicots. Samylina recorded only one modern genus—*Cercidiphyllum* (*Cercidiphyllaceae*)†; but she also pointed out the similarity between the impression she described as *Celastrophyllum kolymensis* and the leaves of the living *Trochodendron aralioides* (*Trochodendraceae*). Although

* However, the correctness of the generic identification of '*Magnolia delgadoi*' is uncertain, while '*Aralia calomorpha*' is clearly not referable to the genus *Aralia*.

† *Sassafras kolymensis* (Krysht.) Baik., mentioned by Samylina, belongs to the organ-genus *Sachalinella* [*S. kolymensis* (Krysht.) Imchanitskaya] of the *Araliaceae*.

Samylina was inclined to accept the Zyrianka fossil flora as Albian, she did not exclude the possibility of its being Aptian. The presence of angiosperms at such a high latitude indicates that these deposits are more probably Albian in age. The remains of small-leaved angiosperms from the Lena (River Lyampeske) basin studied by Kiritchkova and Budantsev (1967) are also apparently of Albian age. The generic composition of this fossil flora is similar to that of the Zyrianka, but more impoverished. The age of the Middle Albian fossil flora of the Kyzylshenskii suite in west Kazakhstan, as determined by Vakhrameev (1952), is more certain. Of the thirty-five species described from this suite fourteen are angiosperms. In the Albian deposits of Primorie, *Aralia lucifera* and *Cissites prodromus* have been found near the town of Suchan, and, on the east coast of the Gulf of Ussuri, *Artocarpidium primigenium*, *Sassafras* (?) *ussuriensis*, *Sapindopsis* cf. *angusta* (impressions of a compound leaf) and remains of dicot fruits. Conifers predominate in the flora and ginkgos are also met with comparatively frequently (Krassilov, 1965). Remains of flowering plants have also been found by Krassilov in the Albian deposits of the Suifun basin near the village of Konstantinovka. They are represented by *Cercidiphyllum suifunense* (leaves and fruits), *Laurophyllum* sp. and other flowering plants, found together with conifers, various ferns and *Bennettitales*.

In North America a whole series of well-studied Albian fossil floras are known, most of which are of certain date and contain various percentages of angiosperms. In the Late Albian fossil floras of Dakota and Aspen (as in the Late Albian flora of Nazaré in Portugal) angiosperms total 85% of the species composition, i.e. they are already dominant in these places.

From the Albian onwards occurrences of angiosperm pollen also became considerably more numerous. Up to the Albian, however, angiosperms appear to be comparatively rare (especially in high latitudes) and, in comparison with gymnosperms and ferns, are everywhere in the minority; only rarely, as in the Funson flora, do they reach 45% of the total flora. They evidently formed only small populations and were represented by a comparatively small number of individuals. A striking characteristic of Neocomian and Aptian fossil floras is the pronounced microphylly of the angiosperms, though in the majority of Albian, and particularly Cenomanian, floras the leaves are considerably larger, many of them being very large indeed. This microphylly of the early angiosperms of Europe, northern Asia and North America is probably due to their having been in the vanguard of northward-moving waves of migration, and thus growing under comparatively unfavourable conditions. We therefore rarely meet with broad-leaved forest elements amongst them.

At the close of the Albian, i.e. towards the middle of the Cretaceous period, one of the most sudden and fundamental transformations of terrestrial plant life occurred, and in the course of a few million years—a very short interval in geological time—the angiosperms came to be widely distributed throughout the world, quickly reaching the Arctic and Antarctic regions. They appear in great diversity of form and quickly become dominant. Most of these Cretaceous angiosperms belong to extant genera, and there are representatives both of more or less primitive forms (*Magnoliaceae*, *Lauraceae* and related families, the genus *Nelumbo*, etc.) and of highly evolved ones (*Fagaceae*, *Moraceae*, *Euphorbiaceae*, *Salicaceae*, *Juglandaceae*, etc.). It is obvious that the angiosperms had previously undergone a protracted period of development during which they had been able to differentiate into many distinct families.

This vast and rapid spread of flowering plants was one of the most important events in the history of life upon the earth. It had a decisive influence on the future of the terrestrial animal world, especially the insects, birds and mammals. In the final reckoning the dominance of flowering plants made possible the appearance of man. The middle of the Cretaceous was therefore the beginning of a new era in the history of life upon our planet.

CHAPTER 11

THE CRETACEOUS EXPANSION—WHY AND WHEREFORE?

How are we to explain the great Middle Cretaceous expansion of the angiosperms, which in a comparatively short time transformed the whole terrestrial plant world; under what conditions did this great floristic revolution occur? The pages of the geological record that are accessible to us not only fail to answer these questions but also set us fresh problems.

In a letter to the famous palaeobotanist Heer, Darwin in 1875 suggested that the angiosperms 'must have been largely developed in some isolated area, whence owing to geographical changes, they at last succeeded in escaping, and spread quickly over the world'. But to this day no one has ever succeeded in putting this plausible hypothesis on a sufficiently concrete factual basis, and in this respect contemporary science has achieved only meagre results. Leaving to the next chapter the question of the region in which the earliest angiosperms most likely developed, we shall now consider the geographical changes that may have facilitated the success of the angiosperms.

During the Jurassic period the development of vegetation went on under comparatively stable conditions, and only the Late Jurassic was marked by marine transgressions and, in places, an increased aridity of climate. The close of the Jurassic was marked by the start of continental uplift, the intensification of mountain-building, the development of arid-climate regions and the general diversification of environmental conditions. These changes of course came about very gradually, and the Cretaceous period was initially very similar to the last stages of the Jurassic; nevertheless, the changes were sufficient to lead gradually to an abrupt transformation of the organic world. The Cretaceous, and particularly the Late Cretaceous, was marked by a much greater diversity of physico-geographical conditions than the Jurassic. For reasons we do not fully understand the plants that had been dominant in the Jurassic period gave way to the angiosperms, either dying out completely, like the *Bennettitales* and most of the *Ginkgoales*, or fading into the background, like the cycads and ferns. Evidently they did not possess sufficient evolutionary plasticity to enable them to produce new forms adapted to the novel conditions of a more diverse and rapidly changing environment. Therefore, first in the mountains and then in the lowlands, the dominant position passed to the angiosperms. Kryshtofovich (1946) was correct in concluding, on the basis of his prolonged studies of the fossil floras of various parts of Eurasia, that the Cretaceous 'replacement of Mesozoic vegetation by Cainozoic took place under the influence of orogenesis, marine transgression and regression, and the resulting widespread climatic changes. Thus, on the one hand, new, unoccupied, land surfaces were created or regions of new environmental conditions established; and, on the other, areas were submerged by the sea and disappeared, or radical alterations in physico-geographical conditions took place which resulted in the total or selective partial extinction of the organisms involved.'

But why was the rapid spread of the angiosperms (and of some other groups of plants) facilitated by these changes of geography? The main reason was no doubt their great evolutionary plasticity and unusual adaptability, as was particularly emphasised by Scott (1911). This plasticity of the angiosperms is clearly shown by their extraordinary diversity; in no other group of plants do we find such vast differences as there are between, for example, a dandelion and a magnolia, or a glasswort and an orchid. All the other members of the Jurassic and Early Cretaceous floras are marked by notably poorer evolutionary plasticity and much less diversity of form. The lower evolutionary plasticity of the Mesozoic gymnosperms is a result especially of the lower level of development of their vascular system, owing to which their leaves, as in present-day cycads, have a xeromorphic structure even in a moist climate. This primitiveness of xylem permitted the gymnosperms to develop only relatively small leaves, or, in the case of those with large pinnate leaves, only a few on each plant and even these with distinctly xeromorphic structure, as in many *Bennettitales* and *Cycadales*. The total area of their photosynthetic surface being thus limited by the under-development of the water-conducting tissues, their chances of success in the struggle for existence were distinctly lowered. The earliest vesselless angiosperms had no advantage in this respect over the Mesozoic gymnosperms, for in them also the photosynthetic surface is not large. The ability to increase significantly the quantity of organic material produced by photosynthesis is evidently closely tied up with the development and improvement of vessels.

According to Voronin (1964), the author of an interesting work on the evolution of the primary structure of the root, 'it is possible to consider that the degree of perfection of the root system played a not inconsiderable role in the success of the angiosperms and, in particular, that the exceptional capacity for root-renewal which they developed determined to a large extent their success under climatic conditions in which an alternation of favourable and unfavourable seasons for vegetative growth prevailed' (p. 157). It is quite possible that this high adaptability of the roots of the flowering plants was also a definite factor in the rapidity of their expansion.

The angiosperms therefore soon became capable of adapting themselves to the sudden changes in physico-geographical conditions which were so characteristic of the Late Jurassic and Early Cretaceous. In particular, they were able to colonise not only mountains but also extensive areas of lowland, and even adapt themselves to the conditions brought about by increasing dryness of climate. As newcomers from the mountains, they quickly conquered large areas of both the tropics and the temperate regions; having reached a high level of organisation and a wide diversity of form in the mountains, they easily emerged as victors in the lowland areas of the earth.

However, perfection of the vascular system, leaves and roots was in itself insufficient to secure for the angiosperms their dominant position on the earth. Vessels are found not only in angiosperms but also in several highly-evolved gymnosperms (*Ephedra*, *Welwitschia* and *Gnetum*), in the genus *Pteridium* (bracken) amongst the ferns, in the horsetails (*Equisetum*), in several species of *Selaginella* and even in the roots of *Marsilea*; but none of these plants plays any major role in present-day vegetation, and the mere presence of vessels alone is of little significance in the conquest of the land. It is true that in the angiosperms a progressive development not only of the xylem elements but of the whole conducting system took place, accompanied by a marked development of the entire metabolism, especially photosynthesis*; but even this was still insufficient to ensure success in the struggle for existence.

An important factor in the rapid development of the angiosperms was evidently their complex interrelationship with the insects, a relationship which is not found, or is only slightly developed, in living gymnosperms, as in *Welwitschia*. The part played by insects in the evolution of the flower was first noted by Müller and the French palaeobotanist Saporta, whose observation met with the warm commendation of Darwin, who in 1877 wrote in a letter to Saporta, 'Your idea that dicotyledonous plants were not developed in force until sucking insects had been evolved seems to me a splendid one. I am surprised that the idea never occurred to me, but this is always the case when one first hears a new and simple explanation of some mysterious phenomenon.' Later, the role of insects in the evolution and development of the angiosperms was particularly emphasised by Arber and Parkin and by the palaeobotanists Seward (1910), Scott (1911) and many others after them. 'It is probable,' wrote Scott, 'that the close relation to insect life has been the chief condition determining the evolution of angiosperms and giving them their supremacy among living vegetation.' Like Saporta and Darwin, Scott attached exceptional importance to the insect pollinators in the evolution of the angiosperms: 'When the angiosperms came in so suddenly, as it seems to us, in Cretaceous times, the whole face of the world was changed, and flowers like those with which we are now familiar everywhere began to

* Teslenko (1967) has recently suggested that the major reason for the rapid transformation of the flora in Mid-Cretaceous times was a sharp decrease in the carbon dioxide content of the atmosphere. According to Teslenko, the early flowering plants, which lived in mountains, were well adapted to the conditions of a low content of carbon dioxide. When in the Mid-Cretaceous the percentage of carbon dioxide decreased and the light-intensity increased, it had a destructive influence upon the Mesozoic plant world. Yet the angiosperms turned out to be the group best adapted to the new gas régime of the atmosphere. This greatly promoted their wide expansion over the whole earth.

appear. This, the greatest change which the kingdom of plants has ever known—almost comparable to the advent of man in the animal record—doubtless chiefly depended on the simultaneous development of the higher forms of insect life' (Scott, 1911). In his opinion, this proposition is not contradicted by the existence of wind-pollinated angiosperms, since it is most probable that wind-pollinated families developed from insect-pollinated ancestors.

Thus it came about that towards the beginning of the Cretaceous period the co-evolution of insects and angiosperms had led to a high development of the mechanism of cross-pollination, and the angiosperms thereafter underwent considerable morphological and ecological diversification. As a result of this co-evolution, insects and angiosperms have become the most numerous groups of organisms and the angiosperms have reached the highest level of organisation in the plant kingdom.

The changes that occurred in the sexual generation also contributed to the success of the angiosperms. Even at the dawn of angiosperm history, the female gametophyte must have undergone profound modification and been strongly reduced and simplified. In the angiosperms we see the most simplified of all female plant gametophytes, consisting of the minimum number of cells. Likewise, the process of fertilisation in the angiosperms is extremely highly specialised and perfected. Here, indeed, is another reason for the evolutionary plasticity of the angiosperms, their ability to advance and vary in very diverse directions.

Thanks to their unparalleled plasticity, as expressed in their inexhaustible morphological and ecological diversity, the angiosperms have become capable of forming many different plant communities, from the simplest to the exceedingly complex. The tropical rain forest is a many-layered community in which even the trees form several layers ('forest upon forest', as Humboldt termed it), and all the layers may be formed exclusively or predominantly of angiosperms. In no group of gymnosperms do we observe this ability to form many-layered communities; herbs are absent in the gymnosperms, and the woody gymnosperms do not exhibit a sufficient variety of life-form to enable them to live together in a many-layered community. In the angiosperms various herbaceous forms arose long ago, and the woody forms are much more diversified than in the gymnosperms, comprising many life-forms quite distinct from one another not only in their light requirements but also in many other physiological and ecological attributes. This diversity enables angiosperms to form a multitude of different communities adapted to survive under the most diverse conditions.

Owing to the differentiation of angiosperm communities into layers both above and below the ground, a very large number of species may exist in a single community. The development of such differentiated communities made up of very varied life-forms enabled the angiosperms to exploit the environment more intensively. That an area may sustain the more life the more diverse the forms that occupy it was put forward by Darwin in 1857, in a letter to Asa Gray.

Complex many-layered communities with a predominance of angiosperms were undoubtedly already in existence when the major expansion of angiosperms began in the Cretaceous period. This circumstance must have assisted their very rapid spread over the surface of the earth. In the struggle for survival between communities of angiosperms and communities of other groups of higher plants, the advantages in most cases must have lain with the former. Although the earliest angiosperms only gradually penetrated communities of gymnosperms and ferns, in time their invasion began to assume more massive proportions. By the middle of the Cretaceous, frontal assaults by entire communities had become possible, and this precipitated their rapid expansion.

In the foregoing discussion we started with the proposition that the rapid expansion of the angiosperms in the Cretaceous was associated with the palaeogeographic changes of that period. The question arises, however, as to whether we can explain such a vast and rapid spread of the angiosperms over the whole earth solely in terms of palaeogeographical changes. Some botanists, e.g. Golenkin (1927), have answered this question in the negative. In Golenkin's opinion, mountain ranges, changes in the configuration of continents and oceans, changes in the direction of winds, and changes in the distribution of precipitation could not alone have produced such a sudden and widespread change in the vegetation of this planet. He considered that such a general change of vegetation must have had some extra-terrestrial cause. 'Thus,' he wrote (1927, p. 64), 'I am inclined to accept the major reason for the advance and success of the angiosperms as lying in some extra-terrestrial and therefore cosmic event. What this event was, I, of course, cannot say. It is possible that up to the Middle Cretaceous the atmosphere was much more cloudy, as some palaeontologists have postulated, and that some change then occurred in the atmosphere that caused the cloud to disappear; or it is possible that the atmosphere itself, perhaps in its upper layers, was less transparent to heat and light, and then for some reason became clearer. Either of these changes would permit an increase in the intensity of the sunlight, with all its associated secondary phenomena.' To the conditions produced by such an increase in light intensity the majority of ferns and gymnosperms proved little adapted; but the angiosperms, being in Golenkin's phrase 'children of the sun', were characterised from the very beginning by the ability to 'endure bright sunlight in the highest degree'. In Golenkin's opinion, these changes in the environment must have caused profound and unusually rapid changes, not only in the plant but also in the animal world. Thus he explained the extinction in the Cretaceous of the giant reptiles by the loss of their habitat and sources of food. Analogous proposals concerning the possible part played by cosmic events in the sudden changes that occurred in the organic world in the Cretaceous period have been put forward by various authors.* All we can say, however, in this respect is that although extra-terrestrial events are by no means excluded as a possible reason for the changes in the plant and animal worlds, at present we still have too little knowledge to accept this conjecture as a working hypothesis.

Whatever the causes of the wide and rapid expansion of the

* See, for example, a very interesting book by the astronomer I. S. Shklovsky entitled *The Universe, Life and the Mind* (1965) in which the extinction of the large Mesozoic reptiles at the end of the Cretaceous period is said to have resulted from a persistent increase in the level of cosmic radiation of the order of tens or perhaps of hundreds of times (p. 63). See also I. S. Shklovsky and Carl Sagan, *Intelligent Life in the Universe*, a translation, annotation and extension of the above (authorised translation by Paula Fern, San Francisco, 1966).

angiosperms in the Cretaceous period were, and whatever the conditions under which it took place, there still remain the questions: Where did the angiosperms come from? In what part of the world was their initial centre of distribution? We attempt to give an answer in the following chapter.

CHAPTER 12

THE CRADLE OF THE FLOWERING PLANTS

We must now consider the geographical location of that 'isolated area' from which, in Darwin's opinion, the angiosperms eventually, as a result of geographical changes, managed to escape and then spread over the whole world. Can we in fact find out what part of the world was the cradle of the flowering plants? It was evidently a region in which they experienced a long period of evolution during which the principal families and many genera were differentiated, and it may also have been their centre of origin; in any case, it was probably not very far from their birthplace.

The Tropical Origin of Flowering Plants

The most diverse opinions can be found in the literature concerning the centre of origin and distribution of the angiosperms. Some authors seek the cradle of the angiosperms in high latitudes and are even in favour of an arctic or antarctic origin, while others seek it in the lower latitudes of the tropics or subtropics.

The most widely held hypothesis (which is still held by some today) has favoured a northern polar origin for the angiosperms. It was first suggested by Heer (1868) and more definitely formulated by Saporta (1877), whose article *L'ancienne Végétation polaire* received immediate support from Hooker in his presidential address to the Royal Society (1879). In the nineteenth century the idea was developed by Asa Gray, Adolf Engler and others, and in the twentieth century it has been supported by Thiselton-Dyer, Seward, Berry, Kuznetsov, Kryshtofovich, Chaney, Just and many other palaeobotanists and phytogeographers. According to this hypothesis the angiosperms originated in the polar region of the Holarctic, and from there spread in successive waves across the whole earth. It is interesting to note that similar views on the importance of the frigid and temperate regions of the earth in the evolution and expansion of various animal groups have been expressed by zoogeographers; one of the most notable exponents of such a view was Mathew, who in his very interesting book *Climate and Evolution* (1915, 1939) maintained that the north temperate zone was the chief centre of evolution and dispersal of the terrestrial vertebrates.

The chief attraction of a postulated polar, or generally northern, origin for the angiosperms is that it affords an easy explanation of the presence of many genera common to the floras of eastern Asia and the Atlantic states of North America. 'The theory of southward migration is the key to the interpretation of the geographical distribution of plants,' wrote Thiselton-Dyer (1909, p. 316). But in fact, this explanation is only apparent; for acceptance of this hypothesis would give rise to a whole series of unsolved phytogeographical problems. The greatest difficulty is the floristic poverty of the Cretaceous and Tertiary floras of the Arctic (and of the Antarctic), which has become clear over the last few decades, and also their secondary and derivative nature. The fanciful legend of the existence in Tertiary times of rich subtropical or even tropical floras in the Arctic has been disproved in the past few decades by the investigations of Edward Berry and others; in particular, Asa Gray's idea that in Tertiary times there had been a rich subtropical flora in the Bering Strait region has proved incorrect. It is now quite clear that climatic conditions in the Arctic have always been unfavourable for the development of any rich and varied vegetation. One cannot but agree with Edwards (1955), in whose opinion the Arctic and Antarctic, 'with their long winters of comparative darkness are most unlikely places for the development of new groups of land plants'. Croizat (1952, p. 545) and van Steenis (1962, p. 272) also have strongly criticised the idea of a northern origin for angiosperms.

Supporters of a polar origin for the angiosperms have often cited the Cretaceous floras of the Arctic, which were once considered to be of Early Cretaceous age. Even if Early Cretaceous angiosperms did exist in the Arctic region, this in itself would prove nothing; but in fact, as an analysis of the oldest known angiosperm fossil floras of the world has shown (Axelrod, 1959), angiosperms did not replace the older Jurassic type of vegetation in the Arctic until Late Cretaceous times. At the beginning of the Cretaceous (in the Neocomian), angiosperms are found only in low latitudes; as Axelrod clearly showed in his diagrams, they are found only south of 45°N and north of 45°S. In both the northern and the southern hemispheres angiosperms reached the higher latitudes only at the end of the Early Cretaceous (in the Albian), and in these latitudes they did not replace the relict Jurassic type of vegetation until the beginning of the Late Cretaceous.

In Europe an angiospermous fruit is known from the Neocomian (Valanginian) of southern France. Leaf-remains first appear in the Aptian deposits of Portugal (Cercal) and Russia (Primorie). As regards the Aptian wood fragments from England (Greensand), Axelrod has suggested that they might have been transported thither from the south. In the more northern Aptian floras of Europe, such as the Klin (near Moscow), and especially in the Greenland floras, angiosperms are entirely absent. In Greenland, angiosperms appear only in the Albian, where in the Kome fossil flora they make up 10% of the flora, whereas in the Albian fossil floras of Portugal they make up about 30% of the earlier Buarcos flora and 85% of the later Nazaré flora. In North America, angiosperms are found in the Neocomian only south of 39°N. The oldest fossil floras containing angiosperms are those of Patuxent (Maryland—Virginia), and Lower Horsetown (California) (Axelrod, 1959). Floras of similar age in latitudes farther north are completely devoid of angiosperms. In Canada angiosperms appear only in the Aptian, and then at first in insignificant numbers, while in Alaska the wholly Jurassic character of the vegetation continues right up to the Albian, and angiosperms are absent from both the Kennicott (*c.* 65° North) and Korvin (70° North) floras. In the late Albian deposits of the Kuk River area in Arctic Alaska (the Kungok Flora) we find two small leaves, apparently of *Nelumbites* (Smiley, 1966). Further south, some floras of Aptian age have up to 45% of angiosperms (Funson of Wyoming), and in the upper Albian floras they begin to predominate. A similar situation can be observed in the development of the Early Cretaceous floras of north-east Asia, and also in the southern hemisphere (Axelrod, 1959). Earlier, Teslenko (1958) came to the conclusion that angiosperms were absent from the Aptian-Albian floras of the central part of the west Siberian lowlands; in his opinion, the colonisation of this area by angiosperms occurred somewhat later, at the beginning of the Late Cretaceous (Cenomanian). As Teslenko pointed out, 'The observed facts do not accord with the hypothesis that the angiosperms originated in the northern part of Angaria and then later spread southwards'. (See also Teslenko, Golbert and Poliakova, 1966.)

The hypothesis of an Arctic (or Antarctic) origin for the angiosperms must therefore be firmly rejected. The facts of palaeobotany which the supporters of this hypothesis were wont to cite have been found to be quite incompatible with it; nor can the present distribution of plants be explained by it. Zoologists are also beginning to discard the idea that the vertebrate animals evolved in, and spread out from, high-latitude centres. In the recent *Zoogeography* of the American zoologist Philip Darlington (1957), the reader will find a long and convincing critique of Mathew's hypothesis of the north temperate zone as an important centre of dispersal for the vertebrates.

As the inadequacy of the long prevalent hypothesis of a polar origin for the angiosperms was revealed, workers turned to seek the cradle of their nativity in the lower latitudes—the tropical and subtropical regions. Therefore in the twentieth century many authors, including Hallier, Diels, Kozo-Poljanski, Irmscher, Golenkin, Bews, Wulff, Camp, Bailey, Axelrod, Fedorov, van Steenis, A. C. Smith and Thorne, came out in favour of the tropical origin of the angiosperms.

Between Assam and Fiji

Hallier (1912b) sought the birthplace of the angiosperms in the basin of the Pacific Ocean on the hypothetical continent of Pacifica, and considered that their ancestors might probably be found in such places as the Andes (from Mexico to Patagonia), the Hawaiian Islands, New Caledonia and New Zealand. However, the very existence of the continent of Pacifica is not supported by the most recent geological investigations of the central parts of the Pacific Ocean,* and therefore Hallier's hypothesis cannot be accepted

* 'We now know that there is no possibility that sunken continents could have

nowadays, at least in its original form. The views of Golenkin (1927), and especially those of Bailey (1949), are somewhat more advanced; they exclude the American continent as a possible region of angiosperm development and do not link the solution of the problem with any risky palaeogeographical hypotheses. Golenkin suggested the southern part of the Angara continent (with Oceania) as the possible place of origin of the angiosperms, but unfortunately gave no reasons for the suggestion. Bailey held an essentially similar view; in his opinion, the present floras of northern Australia, New Guinea, New Caledonia, Fiji and the adjacent areas northwards to southern China have provided, and still provide, the largest number of 'missing links' in the chain of angiosperm phylogeny. For example, of the ten known genera of vesselless woody dicots, five occur in New Caledonia, three of them endemic to that island. Bailey pointed out that these contemporary floras of the southern latitudes have provided us with more structurally primitive dicots than all the fossil floras of northern latitudes. He therefore cast doubt on the hypothesis of a northern origin, and in his later years he addressed to the younger generation of morphologists, palaeobotanists and systematists, interested in the origin of the angiosperms, the following words of fatherly advice: 'Look west, young man, towards the remnants of Gondwanaland.' Bailey's view is essentially a further development of Hallier's original idea and undoubtedly represents a considerable step forward; however, it in its turn needs refinement and development.

The geographical restriction of the most primitive angiosperms to the islands and borders of the Pacific Ocean is most striking. Such primitive families as the *Magnoliaceae* (Fig. 28, p. 144), *Degeneriaceae*, *Himantandraceae*, *Eupomatiaceae*, *Winteraceae* (Fig. 29, p. 146), *Austrobaileyacae*, *Amborellaceae*, *Gomortegaceae*, *Lactoridaceae*, *Calycanthaceae*, *Trochodendraceae*, *Tetracentraceae*, *Lardizabalaceae*, etc., are definitely concentrated about the Pacific basic. However, they are far from evenly distributed; in its number of primitive forms the

been in the area since the Middle Mesozoic and are highly improbable before that time. . . . The findings from exploration of the sea floor have shown that the Pacific Ocean has existed in about its present form and depth since the Cretaceous' (Menard and Hamilton, 1963, pp. 193 and 205). (See also Menard, 1964.)

eastern part of the Pacific basin clearly is inferior to the western part, which embraces a wide area from Assam, Burma, China and Japan to Australia, New Zealand, New Caledonia and the Isles of Fiji. It is here, in eastern and south-eastern Asia, Australasia and Melanesia, that the cradle of the angiosperms must be sought. As the whole pattern of the geographical distributions and connections of the primitive angiosperms (especially the families of the order *Magnoliales*) shows, it was this part of the world which had been, if not the birthplace, then at least the original centre of the widespread Cretaceous expansion of the angiosperms; and this could hardly have been very far from their birthplace. But we cannot as yet say with certainty exactly where between Assam and Fiji the area was in which the angiosperms first developed. Most probably it was situated in what is now south-east Asia,* where the relatively greatest concentration of primitive angiosperms is found. Moreover, the greatest number of the primitive representatives of many families is to be found in south-east Asia. It is therefore quite probable that south-east Asia was the cradle of the angiosperms, or at least was very near to it—see Takhtajan, 1957, 1961; Thorne, 1963; Smith, 1967.

Primitive forms may, of course, persist not only within the area of origin of a taxon but also outside it. Cases also are known of taxa which have completely disappeared from the area of their origin but have persisted in places far removed from it. A classic example is afforded by the evolution of the horse which is, from palaeontological evidence, known to have arisen in North America and to have migrated in the Late Pliocene to the Old World. In America, however, horses died out, while in Eurasia and Africa they achieved considerable diversity. Undoubtedly analogous cases occurred in the history of the angiosperms. Therefore we must treat with great caution any conclusions as to the birthplace of the angiosperms, if they are based solely on existing patterns of distribution; parti-

* According to Dobby (1950), south-east Asia includes the south-eastern part of the Asiatic continent (Burma, Thailand, Indo-China and Malaya) and the Malayan archipelago. In the west, Dobby draws the northern boundary of south-east Asia north of the Tropic of Cancer, including within it upper Burma; for phytogeographical reasons we include also Assam and tropical southern China.

cularly dangerous are conclusions based on the geographical distributions of taxa consisting only of distinct isolated representatives that do not form well-marked phylogenetic series. In this connection, Komarov (1908) wrote: 'General conclusions drawn from the geographical distributions of plants must be based primarily on the distributions of series, not on those of isolated species, for in these there is a much greater element of uncertainty and chance.' This dictum of Komarov undoubtedly holds good also for phylogenetic series of higher taxa—genera, families and orders. Study of such series can give not only a picture of their evolution in time, but also the directions of their movement in space. Thus, when analyses of a sufficiently large number of phylogenetic series all produce similar results, we may consider it highly probable that our conclusions as to the centre of origin, and still more as to centre of dispersal, of a group are near to the truth. The more closely they agree, the more certain the correctness of our conclusions. It is therefore impossible to ignore the fact that the primitive members of many groups of angiosperms, especially of the more primitive groups, themselves clearly tend to occur in south-east Asia and the neighbouring regions of the western Pacific Ocean basin. In the area between Assam and Fiji grow not just isolated primitive types, but whole groups of closely interrelated genera and even families.

We now consider some examples of these. The geographical distribution of the family *Magnoliaceae* (Fig. 28), one of the most primitive in the order *Magnoliales* (see Chapter 7) is one of the clearest examples. The species are concentrated in greatest number in east and south-east Asia; in America they are significantly fewer, and in Africa they are completely absent. But number of species is by no means the only criterion; much more important is the 'phylogenetic geography' of the family. Without exception, all the genera of the family occur in east and south-east Asia, and the most primitive genera and species are confined to that region. Of the twelve genera of the family, only three have representatives in America: *Magnolia*, *Talauma* and *Liriodendron*. The *Magnoliaceae* undoubtedly came to America from Asia, probably by two routes, a northern and a southern, the former taken by *Magnolia* and *Liriodendron*, the latter by *Talauma*. The most primitive species of

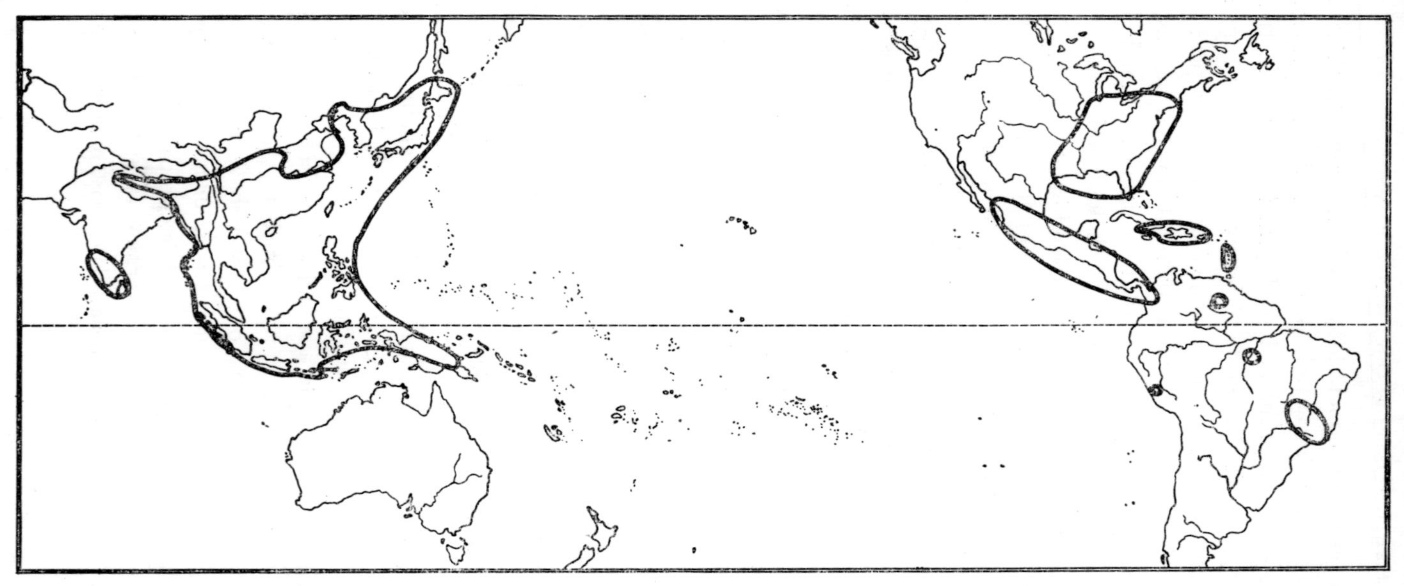

Fig. 28. The distribution of the family *Magnoliaceae*. (After Dandy, 1968.)

Magnolia and *Talauma* are found in Asia; and, what is more, we can observe there almost all the main stages of the evolutionary development of *Magnolia*, from such archaic forms as *M. maingayi*, *M. griffithii* and *M. pterocarpa* to such specialised forms as the Japanese *M. stellata*. The section *Maingola*, which is quite the most primitive in the genus, occurs in Assam and upper Burma and through Indo-China to the Malayan archipelago; it includes *M. griffithii* (Assam and upper Burma), *M. pealiana* and *M. gustavii* (Assam), *M. annamensis* (Annam), *M. maingayi* (Malayan Peninsula and Sarawak), *M. aequinoctialis* (Sumatra) and *M. macklotii* (Sumatra and Java)—see Dandy (1950). As we have seen in Chapter 6, the members of this section are characterised not only by the primitive structure of their reproductive organs and wood, but also by their primitive stomatal apparatus (Baranova, 1962).

The South eastern Asian genus *Manglietia* is very close to *Magnolia*, and in some respects more primitive. Its area extends from the eastern Himalayas, Assam and south China through Thailand and Indo-China to Java. The most primitive representatives of the genus occur in the continental part of south-east Asia.

The families *Degeneriaceae* and *Himantandraceae* stand comparatively close to the *Magnoliaceae*; both are confined to the western part of the Pacific basin; the former is endemic to Fiji, the latter occurs in north-east Australia, New Guinea and the Moluccas. The family *Winteraceae* (Fig. 29) is much more remote from the *Magnoliaceae*, but rivals it in primitiveness. The largest, most widespread and, at the same time, most primitive genus is *Drimys*, the area of which includes the Philippines, Borneo, Celebes, Amboina, New Guinea, eastern Australia, Tasmania, Juan Fernandez, and America from Mexico to Cape Horn. The next in number of species is *Bubbia*, also very primitive, which occurs only in the Old World (New Guinea, Queensland, Lord Howe Island, New Caledonia and Madagascar), as do the four remaining genera. The majority of the species of *Drimys* and *Bubbia* are confined to New Guinea. *Belliolum* is known from New Caledonia and the Solomon Islands, *Pseudowintera* occurs in New Zealand, *Exospermum* and *Zygogynum* in New Caledonia, and *Tetrathalamus* in New Guinea. As Smith (1945a) pointed out, Australasia seems to be the main

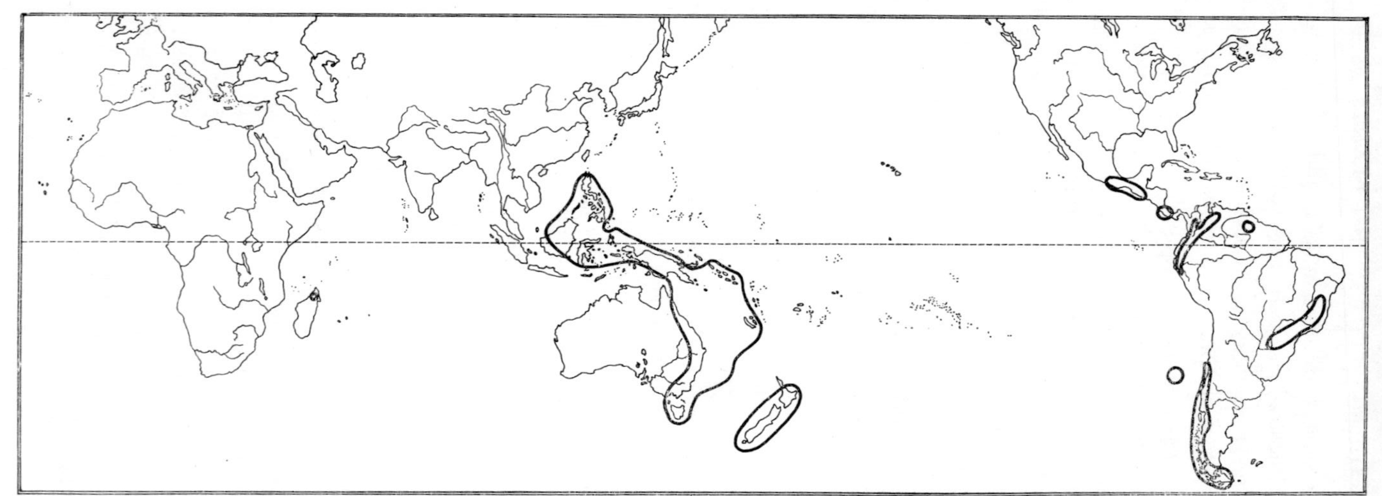

Fig. 29. The distribution of the family *Winteraceae*. (Modified after A. C. Smith, 1943.)

centre of diversity and morphological evolution of the family, and in my opinion is probably also its centre of origin. The primitive section *Tasmannia* of the genus *Drimys*, and almost the whole of the genus *Bubbia*, are in fact confined to Australasia.

The family *Eupomatiaceae* is entirely Australasian; it is characterised (as we saw in Chapter 7) by a series of extremely primitive features. The only genus *Eupomatia* is found in eastern Australia (Queensland, New South Wales and eastern Victoria), and in the southern and eastern parts of New Guinea (Hotchkiss, 1955a).

The family *Annonaceae*, which has attained a higher degree of specialisation than the families already mentioned and is also distinguished by the greater number and diversity of its genera, does not give such a clear picture of its phylogenetic geography. Yet, it is striking that out of 120 genera and 2100 species (Buchheim in Engler's *Syllabus*, 1964), 51 genera and about 950 species are confined to Asia and Australasia, whereas in Africa and Madagascar there are 40 genera with about 450 species, and in the American continent 38 genera and 740 species (Fries, 1959). Thus, Asia together with Australasia is the basic centre of distribution of the *Annonaceae*; if we add to this the fact that the majority of the primitive members in the tribes *Uvarieae* and *Unoneae* are concentrated in south-east Asia and Australasia (as is apparent from Fries's review of the family), then we may safely conclude that this family was originally centred in this part of the world.

The centre of distribution of the family *Myristicaceae* is tropical Asia, from India to Polynesia, where the greatest number of species is concentrated, although the greatest number (7) of genera is found in Africa and Madagascar, and in America there are 5 genera. However, the *Myristicaceae* have been so little studied morphologically and phylogenetically, that at present it is difficult to say exactly which of the 16 genera of this family are the most primitive.

As for the remaining family of the *Magnoliales*, the *Canellaceae*, its distribution lies far outside the bounds of Asia, Australasia and Melanesia. It occurs (as we have seen in Chapter 7) in America, Africa and Madagascar.

Thus, although the most primitive members of the *Magnoliales* are clearly associated in their origin with south-east Asia, Australasia and Melanesia, the situation is less clear as regards the more specialised families, and some of them (e.g. *Canellaceae*) either have now lost any trace of this association or—what is more likely—they arose entirely outside the original centre of angiosperm distribution.

Turning now to the order *Laurales*, we find a very similar situation. Undoubtedly the most primitive families of this order are the *Austrobaileyaceae* and *Amborellaceae*, as we have already seen in Chapter 7. The former occurs in Australia (Queensland), the latter in New Caledonia. The family *Trimeniaceae*, which is also confined to the western part of the Pacific basin, consists of two genera, *Trimenia* and *Piptocalyx*; the former grows in New Guinea, New Caledonia, the Isles of Fiji and the Marquesas, the latter in New Guinea and Australia (New South Wales).*

According to Money, Bailey and Swamy (1950), the most primitive subfamily of the *Monimiaceae* is the *Hortonioideae*, which consists of the single monotypic genus *Hortonia*; this genus is endemic in Ceylon, the flora of which is closely connected in its origin with that of south-east Asia. Considered a little more advanced by these authors is the subfamily *Atherospermatoideae*, in which they include six genera. The most primitive is the endemic New Caledonian genus *Nemuaron*; the next most primitive, *Daphnandra*, is Australian, while *Dryadodaphne* grows in New Guinea and Queensland, *Doryphora* in Australia, *Laurelia* in New Zealand, Chile and Patagonia, and *Atherosperma* (Fig. 20, p. 89) in eastern Australia and Tasmania. The association of this subfamily with the western part of the Pacific basin is obvious: except for one species of *Laurelia*, all the members occur in Australasia or parts of Melanesia. The most primitive members of the large subfamily *Monimioideae*, which consists of 34 genera, have a very similar distribution. One of the most primitive genera, *Hedycaria*, occurs in New Zealand, eastern Australia, the Solomon Islands, New Caledonia, the Fiji Islands and Samoa, the greatest number of species being in

* Hutchinson (1959a, 1964) also includes in the *Trimeniaceae* the genera *Xymalos* (tropical and south Africa) and *Sphenostemon* (New Caledonia and New Guinea), but this is not in agreement with the results of the comparative morphological studies of Money, Bailey and Swamy (1950).

New Caledonia. Another primitive genus, *Levieria*, is found in New Guinea, the Moluccas (Amboina) and Australia (Queensland). From these facts we can conclude that the centre of origin of the *Monimiaceae* was somewhere in the western part of the Pacific basin. But we cannot say the same of the more specialised families of the order *Laurales*—*Calycanthaceae*, *Lauraceae* and *Hernandiaceae*—the present distributions of which give no clear indication of their centres of origin; as for the peculiar family *Gomortegaceae*, it is endemic in Chile, i.e. confined to the eastern part of the Pacific basin.

Much the same applies to the other relatively primitive family of the *Laurales*, the *Chloranthaceae*. The two most primitive genera are *Sarcandra* (Fig. 18, p. 86) and *Chloranthus*, *Sarcandra* having vesselless xylem. Of the two species of *Sarcandra*, one (*S. glabra*) occurs in Japan, the Ryukyu Islands, Taiwan, south China, India, Ceylon, Indo-China and the Malayan Archipelago, the other (*S. hainanensis*) is endemic in Hainan. The genus *Chloranthus* includes about 15 species occurring in east India, Ceylon and in east and south-east Asia. *Ascarina* occurs in Polynesia and New Zealand, *Hedyosmum* in tropical America (40 species) and Hainan (1 species); and *Ascarinopsis* (which is close to *Ascarina*) is endemic in Madagascar (*A. coursii*).

A similar pattern of geographical distribution is shown by the order *Piperales*. The primitive family *Saururaceae* is composed of 5 genera, of which the most primitive is *Saururus* (having scalariform tracheids, apocarpous gynoecia, and monocolpate pollen grains). One of the two species—*S. chinensis*—ranges from central China to Korea, Japan, the Ryukyus, Taiwan, Hainan, North Vietnam (Tonkin) and the Philippines (Luzon); the other—*S. cernuus*—occurs in Atlantic North America. *Gymnotheca*, which is more advanced than *Saururus* but more primitive than the other two genera, occurs in China. The genera *Houtuynia* and *Anemopsis* occur respectively in Asia (from the Himalayas to continental China, Taiwan, the Ryukyus, Japan, Thailand and Indo-China) and in North America, and the monotypic genus *Circaeocarpus* is endemic in south-West China. On the whole, the *Saururaceae* are thus mainly Asiatic (east and south-east), only the genus *Anemopsis* being absent from Asia.

Of the order *Aristolochiales* we shall consider only the most primitive member, the family *Aristolochiaceae*, in which the most primitive genus is *Saruma*. Its flowers have a double perianth, its pollen grains are monocolpate and similar to those of some *Chloranthaceae* (Erdtman, 1952, p. 62), its gynoecium semi-apocarpous and semi-inferior, and its fruit a semisyncarpous multifolliculus. It is monotypic and confined to China. According to Gregory (1956), it is from *Saruma* that the large genus *Asarum* is directly derived; it occurs mainly in the temperate zone of Eurasia and North America. It is interesting that in the species of the sections *Heterotropa* and *Ceratasarum* the ovary is semi-inferior, as in *Saruma*; the species of section *Heterotropa* are found in Japan, and those of *Ceratasarum* in east Asia and North America, with the most primitive species in east Asia. The more advanced genus *Thottea* occurs in west Malesia, and the genus *Apama* (including *Asiphonia* and *Cyclodiscus* sometimes separated as distinct genera) in India, South China, Vietnam and west Malesia. In *Thottea* the flowers are still actinomorphic, and in many species the stamens are still free. The remaining genera of the *Aristolochiaceae*, which are united to form the tribe *Aristolochieae* and are characterised by marks of higher specialisation (zygomorphic perianth and stamens fused with the style to form a column, the gynostegium), are either widely distributed in Eurasia, Africa and America (*Aristolochia*) or confined to tropical America (*Holostylis* and *Euglypha*) or to tropical Africa (*Pararistolochia*).

Also of interest is the order *Illiciales*, which consists of the two families *Illiciaceae* and *Schisandraceae*. The *Illiciaceae*, consisting of one genus *Illicium*, occur in east and south-east Asia and in south-east North America, the greatest concentration of species being in Asia, only two, more advanced, species occurring in America. Of the two genera of the family *Schisandraceae*, one (*Kadsura*) is found only in India, Ceylon and east and south-east Asia, the other (*Schisandra*) mostly in east and south-east Asia, with only one representative in America. The most primitive sections of both genera occur in south China and north Indo-China (see Smith, 1947).

In the order *Ranunculales* (in the narrow sense) the most primitive family is undoubtedly the *Lardizabalaceae*. The most primitive genus of this family is *Decaisnea*, the two species of which are erect

Plate I. *Ficus pubinervis* in tropical rain forest, Peutiang Island, W. Java, Indonesia. (Photo: An. A. Fedorov.)

Plate II. *Ferula badhysi* in the Badkhyz Desert, Turkmenistan, U.S.S.R. (Photo: A. Takhtajan.)

Plate III. Alpine meadows, Mir La, S.E. Tibet, at 3965 m. *Cirsium eriophorum* subsp. *holocephalum* in the foreground. (Photo: Sir George Taylor.)

Plate IV. Mangrove formation, Ceylon. *Rhizophora mucronata* in foreground. (Photo: courtesy of Sir George Taylor.)

Plate V. Candelabra *Euphorbia royleana*, north-east of Dehra Dun, India. (Photo: M. A. Rau.)

Plate VI. *Hornstedtia grandis* (*Zingiberaceae*), Taiping Hills, Malaya. The rhizome grows horizontally about 1 m above the ground. The rhizomes of other species are subterranean. (Photo: R. E. Holttum.)

Plate VII. Carnivorous liana *Nepenthes gracilis*, Singapore Island. (Photo: R. E. Holttum.)

Plate VIII.
Dionaea muscipula, a carnivorous angiosperm (*Droseraceae*).

Plate IX. *Rafflesia arnoldii*, Benkoelen, Kephieng, Indonesia; diameter of flower about 75 cm. (Photo: C. N. A. de Voogd, courtesy of C. G. G. J. van Steenis.)

Plate X. *Hydnora africana*, flower and two flower-buds of this leafless root-parasite, Nkunde-Kisungu, Tanzania. (Photo: A. A. Bullock.)

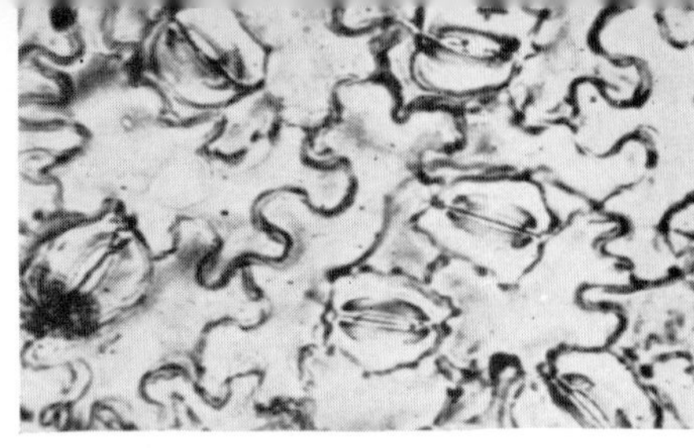

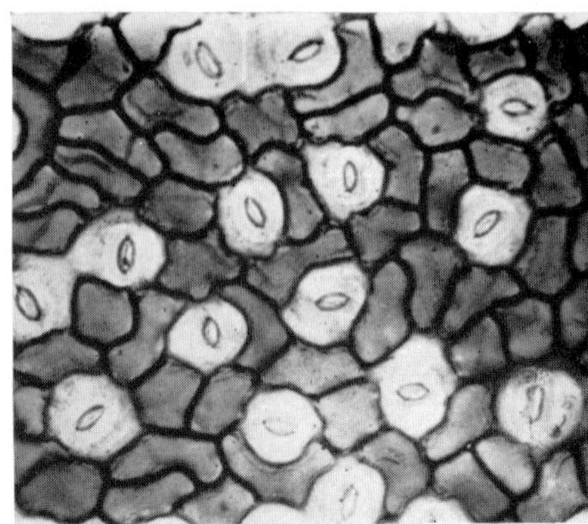

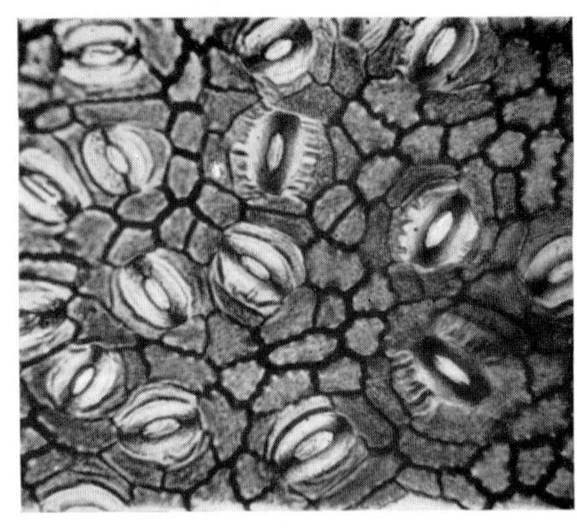

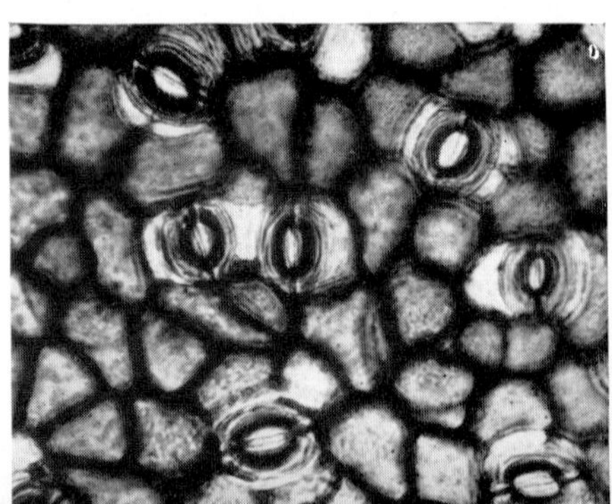

1

2

3

4

Plate XI. Primitive stomata of: *1*, *Manglietia forrestii*; *2*, *Aromadendron nutans*; *3*, *Drimys piperita*; *4*, *Drimys winteri*; all × 240. (Photos: Margarita Baranova.)

Plate XII. Branch of *Austrobaileya maculata* with fruits. (Photo: Webb.)

Plate XIII. Leaf-impressions and fruits of some Early Cretaceous flowering plants: *1*, *Hydrocotylophyllum lusitanicum*; *2*, *Dicotylophyllum cerciforme*; *3*, *Ranunculicarpus quinquecarpellatus* (× 7); *4*, *Crataegites borealis*; *5*, *Nyssidium orientale* (× 1·3); *6*, *Cinnamomoides ievlevii*. (*1*, *2* after Texeira, 1948; *3–6* after V. Samylina.)

shrubs with pinnate leaves and polygamous flowers. It differs from the remaining genera also in the more primitive structure of the vascular system of its stem. *Decaisnea* occurs in the eastern Himalayas and in west China. All the other genera are lianas with unisexual monoecious or dioecious flowers. It is of interest that all the genera with monoecious flowers occur in Asia, from the Himalayas to Japan, *Sinofranchetia* in central China, *Holboellia* in the eastern Himalayas, Assam, China and Tonkin, *Akebia* in continental China, Taiwan, Korea and Japan, *Stauntonia* in Assam, in south China, Hainan, Taiwan, the Ryukyus, Laos, Vietnam, Korea and Japan, and *Parvatia* in Assam, east Bengal, south and west China, and in Tonkin. The two remaining genera—*Boquila* and *Lardizabala*—have dioecious flowers and occur in Chile. From this, the conclusion naturally follows that the *Lardizabalaceae* originated in the western part of the Pacific basin and from there spread to South America—see Croizat (1952, p. 74).

The family *Sargentodoxaceae* stands very close to the *Lardizabalaceae* and consists of one monotypic genus, *Sargentodoxa*, which occurs in west and south-west China, Laos, and also in North Vietnam (Tonkin), where I discovered it in 1960. In its stem anatomy and carpel structure *Sargentodoxa* is more advanced than the majority of the *Lardizabalaceae*, but the carpels are numerous and spirally arranged. The geographical distribution of this peculiar genus supports the conclusion that the west Pacific was the birthplace of the family *Lardizabalaceae*.

The largest family in the *Ranunculales* is the *Menispermaceae*, with as many as 70 genera and about 450 species, and it too on all accounts, had a west Pacific origin. According to Diels (1910), the genus *Pycnarrhena* (tribe *Triclisiae*) stands at the beginning of the *Menispermaceae*; it ranges from the eastern Himalayas and Assam to north-east Australia (Queensland). The allied genera *Macrococulus* and *Haematocarpus* occur respectively in New Guinea and in the eastern Himalayas, Assam and West Malesia. Similar distributions characterise the primitive genera of the majority of the other tribes of the family. For example, in the tribe *Fibraureae* the most primitive genera *Tinomiscium* and *Fibraurea* occur in south-east Asia (*Tinomiscium* also in New Guinea), in the tribe *Tinosporeae* the genus *Aspido-*

carya occurs in the eastern Himalayas, Assam and in south-east Asia, and in the tribe *Menispermeae* ('*Cocculeae*') the genus *Hypserpa* occurs in China, south-east Asia, Melanesia, Polynesia and in north-east Australia.

Without pausing to consider the other families of the order *Ranunculales*, we pass on to one of the most important lines of dicot development—the one beginning with the order *Trochodendrales*, which we have already had a chance to meet in Chapter 7. Both the genera referable to this order occur in Asia, *Trochodendron* in Japan, Korea and Taiwan, and *Tetracentron* in the eastern Himalayas, central and south-west China and upper Burma. Near to the *Trochodendrales* are the families *Cercidiphyllaceae* (China and Japan) and *Eupteleaceae* (Mishmi Hills in the eastern Himalaya, China and Japan), which in my system form the independent orders *Cercidiphyllales* and *Eupteleales*. Near to this order stands the order *Hamamelidales*, the basic family of which is the *Hamamelidaceae*.

The most primitive genera in the *Hamamelidaceae* are evidently *Disanthus* (Fig. 23/*1–5*, p. 99), *Exbucklandia* and *Rhodoleia*, of which *Disanthus* is the most primitive in floral structure and the other two in wood structure (having more than twenty bars in the scalariform perforation plates of the vessels). These three genera also have the most primitive pollen type within the family. Geographically, the montotypic genus *Disanthus* (*D. cercidifolius*) occurs in the mountains of Japan (Honshu and Shikoku), *Exbucklandia* ranges from the eastern Himalayas and Assam to Sumatra, and *Rhodoleia* from south China to Sumatra. Most of the remaining genera of the *Hamamelidaceae* likewise occur in Asia, mainly in the Himalayas and east and south-east Asia. In North America there are three species of *Hamamelis* (the others occur in China and Japan) and also the genus *Fothergilla*, while in Central America there are two species of *Distylium* (the other ten occurring in the Himalayas, Assam, continental China, Taiwan, Korea and Japan). Finally, there is *Trichocladus* in Africa, *Dicoryphe* in Madagascar and the Comores, and *Ostrearia* and *Neostrearia* in north-east Australia. The subfamily *liquidamboroideae* is closely connected to the *Hamamelidoideae* through the genus *Exbucklandia*, and consists of two genera, *Altingia* (Fig. 24, p. 100) and *Liquidambar*. The first of these occurs in Asia, from upper Assam, Japan and China to Java and Sumatra, and the second in Asia Minor, east Asia, Indo-China, North America and Central America. Both the phylogenetic connections and the geographical distribution of this subfamily indicate that its centre of origin was somewhere in the south-east part of the Asian continent.

Without considering the other more specialised members of the *Hamamelidales*, we pass on to some of their more specialised anemophilous derivatives, the *Casuarinales* and their relatives. The order *Casuarinales*, consisting of one family, *Casuarinaceae*, and the single genus, *Casuarina* (Fig. 26/*8–12*, p. 103) possibly originated somewhere in south-east Asia or Australasia. The species of *Casuarina* occur mainly in Australia, but they are also found in New Guinea, New Caledonia, Tasmania, the Malayan Archipelago and the Fiji Islands, while the native range of *C. equisetifolia* is probably from the Seychelles to Australia and Polynesia. The most primitive species of *Casuarina* occur outside Australia (Barlow, 1959); in Barlow's opinion, its birthplace was evidently south-east Asia.

Very great interest attaches to the order *Fagales* (in the strict sense), which consists of the one family *Fagaceae*, widely distributed in the tropics, subtropics and temperate regions of both hemispheres, but absent from tropical and south Africa. According to Hjelmqvist (1958), in its reproductive characters the genus *Lithocarpus* is the most primitive type in the family *Fagaceae*, while according to Forman (1966) *Lithocarpus* and *Chrysolepis* have the most primitive type of cupule. The members of the genus *Lithocarpus* occur in the sub-tropics and parts of the tropics of east and south-east Asia, and the monotypic genus *Chrysolepis* occurs in North America. However, in vegetative anatomy the most primitive genera are *Fagus* and *Nothofagus* (including *Trisyngyne*) (Shimaji, 1962). *Fagus*, that typical representative of the temperate flora of the northern hemisphere, is almost certainly eastern Asian in origin. It is in east Asia that the most primitive and basic types of beech are centred, while the Euro-Caucasian and North American species are derivative in character. Rather closely connected phylogenetically with *Fagus* is the genus *Nothofagus*, which occurs in New Zealand, south-east Australia, Tasmania, extratropical South America, New Caledonia and north-west New Guinea; but this genus is notably more specialised than the beeches, as is shown not only by the structure of its flowers, inflorescences and cupules but also by its wood anatomy, cotyledon morphology and in particular its pollen grains. In Erdtman's (1952, p. 177) opinion, the more or less colpoid apertures of *Nothofagus* pollen grains probably developed from apertures of the *Fagus* type. In general, the genus *Nothofagus* represents a further evolutionary development of the *Fagus* type. There is therefore every reason to believe that the basic prototypes of these two closely related genera reached the southern hemisphere by way of Malesia and Australasia.

The family *Betulaceae* is usually included in the order *Fagales*, but, following Nakai (1943) and Hjelmqvist (1948), we prefer to refer it to the separate order *Betulales*. *Alnus* is the most primitive genus of the family, and within *Alnus*, *Clethropsis* is the most primitive section, the species of which occur in the Himalayas, Assam and China. The other primitive species of *Alnus*, and likewise those of *Betula*, are also confined to the Himalayas and east Asia. A similar situation is observed in the family *Corylaceae*, all four genera of which occur in east Asia, where their most primitive species are found.

The order *Balanopales*, consisting of one family *Balanopaceae* with one genus *Balanops*, occurs mostly in New Caledonia but also in Australia (Queensland), New Hebrides and Fiji.

The *Myricales* (with one family *Myricaceae*) are widely distributed in both hemispheres, but the most primitive representatives occur in east and south-east Asia. The most primitive inflorescence structure, for example, is characteristic of *Myrica rubra* (*M. nagi*) (Assam, China, Korea and Japan) and *M. javanica* (Hjelmqvist, 1948).

The two families *Rhoipteleaceae* and *Juglandaceae* together make up the order *Juglandales*. The family *Rhoipteleaceae* is the more primitive. In many respects it is close to the hypothetical intermediate linking the *Hamamelidales* and *Juglandales*. It consists of one monotypic genus *Rhoiptelea*, which occurs in south-west China and North Vietnam (Tonkin).

Amongst the '*Amentiferae*' only the *Leitneriales* are not represented in Asia at the present day. This highly specialised order (consisting of the family *Leitneriaceae*, with one monotypic genus *Leitneria*) is also probably derived from the *Hamamelidales* or their immediate ancestors. It is confined to the south-eastern states of North America, but fossil remains of *Leitneria* have been found in the Upper Tertiary deposits of west Transcaucasia and west Siberia.

Finally we consider the geographical distribution of the order *Dilleniales*, which as a 'nodus' of phylogenetic connections is one of the most important of the dicotyledons. Two families are placed in the *Dilleniales*—*Dilleniaceae* and *Crossosomataceae* (Hutchinson, 1926; Takhtajan, 1967). The larger, and in several respects the more primitive, is the *Dilleniaceae*, which occurs in the tropical regions of both hemispheres, especially in Australasia and tropical America. The most primitive group is evidently the tribe *Dillenieae* (Ozenda, 1949, p. 95), which contains the genera *Dillenia* (including *Wormia*), *Schumacheria* and *Didesmandra*. *Dillenia* extends from Madagascar and the Seychelles to Fiji; to the north it reaches the southern slopes of the Himalayas, and to the south, Australia (one species in Queensland). The greatest number of species is found in south-east Asia, especially in Malesia. The oligotypic genus *Schumacheria* is confined to Ceylon, and the monotypic *Didesmandra* is a Bornean endemic. The largest tribe of the family, the *Hibbertieae*, occurs mainly in Australia and Tasmania, but more than twenty species of *Hibbertia* are found in New Caledonia, one in the Fiji Islands, two in east Malesia and one in Madagascar (see Hoogland in van Steenis, 1951). It is interesting that what is apparently the most primitive section of *Hibbertia* occurs in New Caledonia. The distribution area of the tribe *Acrotremeae*, which consists of the single genus *Acrotrema*, includes Ceylon, part of India (Deccan), lower Burma, peninsular Thailand and Malacca. The largest area is covered by the tribe *Tetracereae*, which Gilg and Werdermann (1925) place at the beginning of the *Dilleniaceae* and Ozenda (1949), with, I feel, more justification, at the end. *Tetracera* itself is pantropical, and *Davilla*, *Curatella* and *Doliocarpus* neotropical. The geographical distribution of the family as a whole indicates that its centre of origin was most probably situated in the western part of the Pacific basin, between India and New Caledonia.

The second member of the *Dilleniales*, the family *Crossosomata-*

ceae (with a single genus *Crossosoma*), occurs in America (southern California, Santa Catalina Island, St Clement's Island, Guadaloupe, Arizona and north-west Mexico), but, as Eames (1961, p. 435) pointed out, the *Crossosomataceae* are the most advanced members of the *Dilleniales*, being characterised in particular by a more specialised wood structure.

The order *Paeoniales*, with one family *Paeoniaceae*, is quite near to the *Dilleniales*. Corner (1946) first pointed out that the family *Paeoniaceae* (with one genus *Paeonia*) was much closer to the *Dilleniaceae* than to the *Ranunculaceae*, and this has been confirmed by later investigations—see Eames (1961). Following Nakai (1949), I refer it to the separate order *Paeoniales*. *Paeonia* is quite widely distributed, in Europe, the Mediterranean, temperate and subtropical Asia, and the western states of North America. The most primitive arborescent species of *Paeonia*, forming the section *Moutan*, are centred in western China (Kansu, Szechuan, and Shensi).

An analysis of the present distributions of the more primitive living angiosperms thus leads to the conclusion that the original centre of distribution of the angiosperms was situated somewhere between eastern India and Polynesia, i.e. in an area on the west side of the Pacific Ocean basin and including a part of the Indian Ocean basin. In all probability it was the part of the world that today is south-east Asia. This conclusion is probably right, though of course it cannot claim to be certain.

Centre of Origin, or Centre of Survival?

The reader may already have noticed that the theory of the south-east Asian origin of the angiosperms, as elaborated here, has its own 'Achilles' heel'. Its weak spot is the absence of any direct evidence in the geological record. All the conclusions are based on indirect evidence obtained from the study of living angiosperms; but can *any* really valid conclusions be drawn as to the centre of origin or initial distribution of the angiosperms solely on the basis of their present distribution patterns? Are we not taking a mere refugium, or centre of survival, of primitive angiosperms in the 'paradise garden' of the islands and peninsulas of south-east Asia for the cradle of the angiosperms and their initial distribution centre? This question requires to be answered.

Unfortunately the location of the birthplace and primary centre of the angiosperms cannot be ascertained from our knowledge of palaeobotany. As we know, fully authentic remains of angiosperms are found only from the Early Cretaceous onwards, and the Early Cretaceous occurrences of angiosperms are so fragmentary and so few that they can tell us nothing about the location of the centre from which they initially spread. All we can say, on the basis of our less than meagre information, is that the angiosperms spread from low latitudes to high, and not in the opposite direction. The later history of the great angiosperm expansion, which began approximately at the middle of the Cretaceous period, also gives us little help towards solving the problem before us. As for the angiosperms of the Tertiary period, their history is undoubtedly full of absorbing interest and can provide information invaluable for the answering of numerous questions concerning their present distribution, but our knowledge of them is still very inadequate and covers only a comparatively small part of the globe. They are too fragmentary and in many cases of too uncertain identity to serve as material for phylogenetic and phytogeographical analysis in the same way as living angiosperms living today. Although the geographical distribution of living angiosperms is very different from that of the Tertiary angiosperms, and even more so from that of the Cretaceous ones, the living angiosperms do in some cases exhibit sufficiently complete phylogenetic series, and their patterns of distribution can be studied so much more fully, so that they are of immeasurably greater significance for the solution of the problems before us. It is sufficient to note that the vast majority of the genera we have mentioned are unknown in the fossil state, and the ones which do appear as fossils in Cretaceous or Tertiary deposits are usually known from only a few geographically isolated localities; and this gives us no indication of their actual distribution areas. For our purposes, the greatest interest lies in the primitive groups of existing angiosperms, and we can judge their phylogenetic connections solely from the study of these. However one-sided this approach may be by itself, in this instance it is unfortunately the only one possible. We can see what is in many respects an analogous situation in the attempts to solve ethnogenetic problems by the method of comparative linguistics—for example, the problem of the centre of origin of the Indo-Europeans. If the present distribution of the angiosperms over the face of the earth is a direct result of their previous historical development, then the present might well prove the key to the past. Everything depends on the manner and the method of the enquiry. The employment of phylogenetic systematics is especially important in this case. Diels (1908), in his excellent outline of plant geography, emphasised the importance of the phylogenetic method in phytogeography: 'It is true that the use of the phylogenetic method requires much care and effort,' he wrote, 'but its advantages are enormous.' The employment of this method in phytogeography entails 'the study of successive distribution areas and their interrelationships in an order determined by the relevant family tree' (Kozo-Poljanski, 1922, p. 154). Using the phylogenetic method, in this way we can project, as it were, the family tree on to the geographical background. Inasmuch as the direction of evolution of taxa is shown by their family trees, it will be projected as the direction of their dispersal.

The extensive region between Assam and Fiji cannot be simply a 'refugium' of ancient angiosperms, or, in other words, merely a centre of their survival. In fact, during the Mesozoic and Cainozoic climatic and general physicogeographical conditions were no more stable in the area of east and south-east Asia, Australasia and Melanesia than in tropical America or even in tropical Africa. In spite of that, the extremely rich flora of tropical America is significantly poorer in primitive angiosperms than the flora of the western part of the Pacific basic, and the flora of tropical Africa is almost devoid of them. It is difficult to explain why no members of the *Magnoliaceae*, *Winteraceae*, *Himantandraceae*, *Degeneriaceae* or any other archaic family have survived in the African flora, if they had existed there previously. Of the families of the *Magnoliales*, only the *Canellaceae* and the *Annonaceae* are represented in Africa. In Madagascar there are rather more representatives of primitive families than in Africa, as, besides the *Canellaceae* and *Annonaceae*, we also find *Bubbia perrieri* (*Winteraceae*) and *Ascarinopsis coursii* (*Chlorantha-*

ceae), as well as the endemic family *Didymelaceae* with a single primitive carpel and primitive wood anatomy, placed by me in a separate order, the *Didymelales*. Although primitive angiosperms are appreciably more numerous in America than in Africa and Madagascar, even in America they are still far fewer and, as a rule, more specialised than in east and south-east Asia. The American members of the *Magnoliaceae* and *Winteraceae*, for example, are markedly inferior in number and diversity to those of the western Pacific and are representatives of the less primitive forms (see Chapter 7). It is exceedingly unlikely that the concentration of so large a number of primitive angiosperms in east and south-east Asia, Australasia and Melanesia can be explained solely by the presence in this region of optimum conditions for their survival. The variation in palaeogeographical conditions in this region has been not less, but significantly greater, than in tropical America and in tropical Africa.

In Assam, upper Burma, Yunnan, in the mountain regions of Thailand, Laos, and Vietnam, and in Malesia and Melanesia grow several very primitive *Magnoliales* and also many other 'living fossils'. When one considers this striking concentration of primitive angiosperms, amongst which are not a few that are 'missing links', and has visited some of these lands for oneself, one is forced to the conclusion that this region of the world is a fragment of the ancient area which was first colonised by the angiosperms. Admittedly, this conclusion is still far from being proven, but it is very likely that the initial centre of angiosperm expansion (and perhaps also their centre of origin) was situated somewhere in or near this region.

However, in addition to this striking concentration of primitive forms which makes south-east Asia (and the adjoining parts of north-east India, east Asia, and Melanesia) a veritable 'land of living fossils', the idea expounded here has another very important foundation, to which other authors have already given their attention. Thus Golenkin (1927, p. 85) wrote: 'The southern part of the Angara continent, together with Oceania, also provides us in this respect with a basis on which we may explain the possible migration of angiosperms from one continent to another.' A similar idea was put forward by Croizat (1952, p. 109), who emphasised the importance of Malesia as a centre of phytogeographical connections.

'Malaysia is indeed one of the most important phytogeographic zones of the world, and all five continents can easily be reached from its boundaries.' All this is undoubtedly true and may be considered as weighty evidence in support of the hypothesis that the cradle of the angiosperms was situated somewhere between Assam and Fiji, most probably in the area corresponding to what is now south-east Asia. From there they had the opportunity to spread over the entire northern hemisphere. Temperate and subtropical elements were able to reach North America very early by way of the North Pacific track across the Bering-Aleutian area. From south-east Asia angiosperms could well have colonised also the territories of the southern hemisphere. Continental south-east Asia was in fact linked to the Malayan Archipelago for a long period of time, the mountains of the latter having formed a prolongation of the Burma-Yunnan highlands, and, through Malesia, also to New Caledonia, the Fiji Islands and Tonga, north-east Australia, Tasmania, New Zealand, Chatham Island and Antarctica. The intense orogenic folding that took place at the close of the Jurassic and in the Early Cretaceous created a reliable migratory route through the Malayan Archipelago and Australasia to Antarctica. Through Antarctica, and also probably by a somewhat more northerly route, connection was made in the Cretaceous period with the Andes of Chile. Scottsberg (1925) and van Steenis (1962) consider that in the Cretaceous and Early Tertiary the subantarctic temperate flora occupied a rather wide land-bridge which formed an extension of the present Antarctic continent between 40° and 60° S, connecting south-east Australia, Tasmania, New Zealand and Patagonia, in the west extending northward also to Melanesia (Lord Howe Rise and Kermadec Ridge). 'It is not impossible that the northern shores of this bridge could even serve a transpacific exchange of a substantial part of the now southern Pacific subtropical element,' says van Steenis (1962, p. 346). But, as Smith (1963, p. 248) rightly points out, 'To account for known distribution the Tertiary and Cretaceous connections need not have been solid "bridges"; they could have been insular chains, in which connection a thorough search for guyots in the region will be highly important.' In Zimmerman's (1963, p. 479) opinion, 'It may be demonstrated in the future that

Antarctica was never actually joined to other continents with the possible exception of South America, but that chains of steppingstone islands may have allowed biotal penetration, just as we now see this feature in action in the mid-Pacific.' Axelrod (1960, p. 271) also reached the conclusion, that the unity of the Antarcto-Tertiary flora was due to migration over extensive archipelagos which have since subsided. 'These connections were probably between Antarctica and South America along the Scotia arc and between Antarctica and Tasman area along the Macquarie Swell and the New Zealand Plateau.' It is most probably by such a route that representatives of the families *Winteraceae* and *Lardizabalaceae*, and various genera like *Eucryphia* (*Eucryphiaceae*), *Nothofagus* (*Fagaceae*) and *Hebe* (*Scrophulariaceae*) reached America. In this connection, the reader should be reminded of the two species of *Laurelia* (*Monimiaceae*), one of which grows in New Zealand and the other in Chile. Also in the Cretaceous period, south-east Asia was connected, by way of India, Ceylon and the ancient land of Lemuria (of which Madagascar, the Seychelles and the Mascarene Islands are remnants), with Africa, which, in its turn, must have had an ancient (probably also Cretaceous) link with South America, most probably through a chain of steppingstone islands. It has been suggested, for example, that the *Dipterocarpaceae* (Bancroft, 1933) migrated from Asia to Africa by way of Lemuria; and this is supported by the presence of *Vateriopsis seychellarum* in the Seychelles and *Monotes madagascariensis* in Madagascar (see also Wild, 1965). It is possible that the *Hernandiaceae* reached Africa by the same route; they also reached tropical America. South America thus had at least two links with south-east Asia, one through Africa and one across the broad Antarctic land-bridge or chains of 'steppingstone islands'; what is more, the disjunct amphi-transpacific distributions of many tropical families and genera cannot be explained unless one supposes that there existed in the Cretaceous period a link between tropical Asia and tropical America across some fairly large and widespread archipelago (van Steenis, 1962, 1963). At the junction of all these links stood south-east Asia. Therefore the study of the flora of this region (especially its relict forms) is of paramount importance for phylogenetic angiosperm geography. It must be the centre of attention for evolutionary morphologists, phytogeographers and systematists who are interested in the problem of angiosperm evolution; the attention of palaeobotanists must also be directed there.

Darlington, in his book on zoogeography (mentioned earlier) paints a somewhat similar picture of the basic migration paths of terrrestrial and freshwater vertebrates. He concludes that the region from which various vertebrate groups, such as the freshwater fishes, amphibia, reptiles and mammals, spread out was the tropics of the Old World. In his opinion one of the paths led to temperate south Africa; however, the majority of south African vertebrates show a clear affinity with those of tropical Africa. A second major path led from tropical Asia to Australia and New Zealand. The chief barrier on this route was salt sea-water, and therefore the majority of freshwater fishes got no further than Java and Borneo. The *Amphibia* reached Australia several times, and on one occasion New Zealand, but the journey was evidently accomplished only sporadically and with difficulty. The lizards, however, easily reached Australia and also New Zealand, but the terrestrial genus *Testudo* got only as far as the Moluccas. Many birds reached Australia, but only a few reached New Zealand. Of the mammals, apart from numerous bats, comparatively few reached Australia, while New Zealand was reached by only two species of bats. As Darlington remarked, the diversity of the different classes of vertebrates in Australia and New Zealand is approximately proportional to the ability of the animals concerned to cross salt water. Finally, a third major path ran from the Old-World tropics in an arc through temperate Eurasia and North America to tropical Central and South America. Darlington's hypothesis has been given here in a very sketchy outline, but one that will suffice for our purpose. The similarity of the independently derived concepts of phyto- and zoogeography is evident. It only remains to note that his initial centre of distribution for the terrestrial and freshwater vertebrates covers a very large area—the Old-World tropics; and this is quite probable, if one bears in mind the systematic diversity of the group and the different times at which its different classes originated. For the angiosperms the corresponding area was naturally much smaller, but it is highly significant that their postulated cradle, or at least their initial distribution centre, appears in all probability to be a part of the same general area that was by all accounts the centre from which not only the mammals but also the reptiles, amphibians and freshwater fishes first dispersed.

CHAPTER 13

THE ECOLOGICAL EVOLUTION OF FLOWERING PLANTS AND THE ORIGIN OF THE TEMPERATE FLORA

In 1894 the eminent Russian phytogeographer Krassnov put forward the theory that the vegetation of the cooler regions of the world had been formed from the tropical flora by natural selection of those elements which had the ability to grow under low-temperature conditions. This idea, as it applied to the angiosperms, was later expressed by Hallier, Diels, Golenkin, Bews, Fedorov, Thorne and others. The majority of present-day writers consider the angiosperms to be 'children of the tropical sun' and look for their origin within the tropics. The idea of a tropical origin for the angiosperms was developed in particular by the palaeobotanist Axelrod (1952, 1960), who came to the conclusion (1952, p. 44) that 'the tropical uplands probably were the principal reservoir of early, primary angiosperm evolution.' Study of the ecology and geography of primitive angiosperms living today leads inescapably to the conclusion that the angiosperms originated, and for a rather prolonged period evolved, under montane tropical conditions. It seems very unlikely that they could have arisen and undergone their first stages of evolution in the climate of the lowland tropical rain forest, with its extreme constancy of temperature throughout the year. More probably, they developed in the climate of tropical montane forests, which are characterised by a diversity of environmental conditions and by greater geographical and temporal variation. In the forests of tropical mountains we pass from zones dominated by a purely tropical flora to zones with mixed floras containing a large proportion of subtropical and even temperate elements, and the higher we go the more numerous the latter become. The systematic composition of the tropical montane forests of Asia is of especial interest to us.

At the beginning of 1958 I had the opportunity of visiting a well-preserved piece of montane rain forest in the southern part of tropical Yunnan, about 20 km west of Tin-pin. The forest is situated at an altitude of 2200 m on peculiar yellow-brown soils. Because of the high atmospheric humidity all the trees are densely sheathed with bryophytes and innumerable other epiphytes. In its composition the forest differs sharply from the tropical vegetation at lower altitudes. The characteristic plants of the tall tree layer (stratum) are species of *Lithocarpus*, *Acer*, *Manglietia*, mighty specimens of *Phoebe sheareri*, *Actinodaphne* etc. As the eye is caught by the primitive *Manglietia wangii* in the higher layer, so in the lower layer *Illicium majus* and *Tetracentron sinense* engage the attention. In a comparatively small area one finds together several primitive angiosperms belonging to various orders; *Litsea lancifolia* (*Lauraceae*), one of the most noticeable plants of the upper layer, occurs also in the lower. Some of the most interesting plants in the lower layer are *Rehderodendron* (*Styracaceae*) and *Camellia oleifera*. The third layer of the forest, which is not always clearly demarcated from the second, proves no less interesting; the bamboo *Sinocalamus* is very abundant, together with the araliad *Brassaiopsis*, *Hydrangea aspera s.l.* (*Hydrangeaceae*) and *Dichroa febrifuga* of the same family, *Skimmia arborescens* (*Rutaceae*) and such interesting plants as *Plagiopetalum henryi* (*Melastomataceae*) and *Daphniphyllum*. There are also lianas in the greatest diversity and abundance. Some of the commonest are species of *Actinidia* and the herbaceous *Streptolirion volubile* (*Commelinaceae*). The most vigorous are *Holboellia latifolia* (*Lardizabalaceae*) and *Schisandra neglecta*, both members of primitive groups of angiosperms. The following lianas may also be found: *Kadsura* (*Schisandraceae*), the legume *Derris scabricaulis*, *Celastrus stylosus*, *Tetrastigma obtectum*, *Rubus malifolius*, *Polygonum dielsii*, *Gomphogyne cissiformis* (*Cucurbitaceae*), *Dactylocapnos* (*Fumariaceae*) and *Asidopterys obcordata* (*Malpighiaceae*) amongst others. The numerous epiphytes are just as diverse; we find amongst them, besides a multitude of bryophytes and ferns, such different angiosperms as *Aeschynanthus buxifolius* (*Gesneriaceae*), *Polygonatum*, *Rhododendron*, the bilberries *Vaccinium petelotii* and *V. chapaiense*, and even *Sorbus granulosa*. The terrestrial herbs are also very varied; together with many different ferns grow species of *Elatostema*, *Impatiens*, *Scutellaria*, *Polygonum*, *Pellionia*, *Begonia*, *Piper*, *Viola*, *Ophiopogon* and *Musa* (confined to the valleys) amongst others. The extraordinarily diverse floristic composition of this mainly subtropical forest, in which many tropical and Holarctic genera are found growing side by side is quite remarkable. But its most striking feature is the abundance of primitive forms as regards both the angiosperms as a whole and individual orders and families. I was also able in 1960 to study analogous forests in the mountains of North Vietnam, near Cha-pa (Tonkin) and in 1966 in the Khasi Hills (Assam), which likewise contain many primitive forms.

If we now consider lowland tropical vegetation, for example that of the tropical rain forest, we find the proportion of primitive forms is notably smaller; despite the fact that the tropical lowlands are characterised by the greatest diversity of species, genera, families and life-forms amongst the angiosperms, they contain significantly fewer primitive and phylogenetically intermediate forms than the mountain forest of tropical Asia and the subtropical forests of the eastern Himalayas, Assam, upper Burma, North Vietnam, China and Japan. It is in this region—the zone of contact and overlapping of the tropics and subtropics of Asia—that we find the greatest number of relict archaic 'living fossils', which have persisted since the earliest epoch of angiosperm development (even in spite of the higher rates of speciation in the mountains).

As we have already seen in Chapter 4, the angiosperms in all probability arose in the mountains, i.e. under the most diverse and variable environmental conditions. As a result of this variability of environment and the great adaptability of the angiosperms themselves, their evolution proceeded from the very beginning along the lines of adaptive radiation, and they fanned out in many different ecological and morphological directions. If the early angiosperms did grow in conditions analogous to those in which contemporary tropical montane forests exist, then their tempo of evolution must have been very fast, and their adaptive radiation rapid. This is a result not only of the variability of the environment but also of the high 'competition pressure' of life in the tropics. Because one small area of tropical soil can support an extremely large number of different plant species, the number of individuals of each species is rarely very great. This has been noted by various authors, e.g. Richards (1957); Fedorov (1966), who has studied especially the tropical vegetation of south-east Asia, also emphasises the paucity of individuals of any one species in the tropical rain forests.

In passing from the lowland rain forest to the montane forest, we find the number of individuals of many species increases, but the populations often still remain comparatively small, and, what is especially important, the degree of their spatial isolation usually increases. As we saw in Chapter 4, the existence of small, more or less isolated, populations facilitates genetic drift, which can, in favourable circumstances, lead to rapid evolution. Viewed in this light, the strikingly high local endemism of the tropical rain forests (especially in south-east Asia) is understandable (Merrill, 1946, p. 81). According to Dubinin (1940), 'genetico-automatic processes' (genetic drift) have played a very considerable role in the origin of the specific richness and high local endemism of the tropical fauna. 'It is quite possible,' he wrote, 'that the comparative smallness of the populations of tropical forms constitutes the basis of the rapid origin of the great diversity of races and species, both as to adaptive features and as to various indifferent or neutral characteristics which have become fixed by genetico-automatic processes' (p. 297). In his interesting paper on speciation in the tropical rain forest, Fedorov (1966) has recently come to similar conclusions. I am quite certain that the high local endemism of plant species is largely due to genetic drift, the evolutionary significance of which is still underestimated by the majority of botanists. The early angiosperms, which at first were but a scattering in mountain forests consisting of ferns and gymnosperms, must have existed as small isolated populations, and therefore must have been highly subject to genetic drift (Takhtajan, 1947, 1956, 1961). If, therefore, the angiosperms did in fact originate in the mountains of the tropics, then their rapid ecological and morphological differentiation must have begun in the very earliest stages of their evolution. Their basic life-forms and family distinctions were evidently developed within the bounds of their initial centre of distribution.

The majority of living primitive angiosperms occur in areas with a subequatorial monsoon climate, which is characterised by an alternation of hot dry and moist rainy seasons. A markedly monsoon climate prevails in Hindustan (where it affects the Himalayas), in east and south-east Asia north of a line running through the northern part of Malaya and the island of Mindanao, and to the east and south of central Java. In these regions at least one month of the year has less than 60 mm of precipitation. It seems to me there are some grounds for considering that the angiosperms first developed in a subequatorial monsoon climate. It is probable that from the very beginning they had evolved a fairly high degree of adaptability to variations in atmospheric and soil moisture, certainly higher than most gymnosperms and ferns. The improvement of the vascular system of the axial organs of the angiosperms also was possibly a result of the ecological conditions of a seasonal climatic rhythm.

Even in the very earliest stages of their ecological evolution the angiosperms must have become adapted to the various altitudinal belts, with their various climatic and edaphic conditions; and therefore even in their place of origin the angiosperms must have evolved their basic life-forms and some of their basic types of plant community. It seems highly probable that deciduous forms of woody angiosperms had already appeared. As early as 1894 Krassnov pointed out that 'our winter-deciduous forests are not primarily the result of adaptation by the trees to the austerity of winter, which is shown in other ways in the structure of their buds and bark. The periodicity of growth is evidently an attribute which had already been developed in the tropics and later, on the strength of its utility, became usual in the genera and families of our climate.' It is natural to suppose that such periodicity of growth arose at first under the influence of a monsoon climate as an adaptation for survival during the dry season. Quite independently to the same idea came Holttum (1953), who wrote that 'the deciduous habit may have evolved in the tropics, and permitted such trees to spread from the equatorial belt to more seasonal regions'. This idea was later developed by Fedorov (1957), Takhtajan (1957b, 1961) and especially by Axelrod (1966). According to Axelrod, the deciduous habit presumably originated in the lower middle latitudes of the tropical zone and may have been an adaptation to moderate drought in the cooler part of the year (p. 13). It is possible that the earliest angiosperms were also characterised by several other features of xeromorphic organisation which later served as a basis for the xerophilous directions of their ecological evolution.

Before the start of their widespread expansion in the Cretaceous period, the angiosperms no doubt possessed a considerable degree of ecological and phytosociological diversity. When therefore, they were given the opportunity of widespread expansion, they migrated differentially along different geographical and ecological pathways and underwent further transformation en route. While the thermophilous elements won elbow-room in the tropics, the more cold-tolerant deciduous forms made their way along the mountain chains into the cool-temperate and frigid zones, where they rapidly colonised the lowlands. Consequently, the basic elements of the moderately thermophilous and psychrophilous floras had their origin not within the cool-temperate and frigid zones, but far outside their limits. The basic material as such, of the temperate and psychrophilous floras originated in the high mountains of the ancient homeland of the angiosperms, whence growing in diversity and richness they migrated along appropriate ecological channels and migrational pathways. As a result, their original vertical zonation was transformed, as it were, into a secondary latitudinal one (Takhtajan, 1957b).

Especial interest attaches particularly to the temperate flora of Eurasia. The very rich flora of subtropical China (especially Yunnan) and the adjoining areas (eastern Himalayas, Assam, upper Burma, North Vietnam and south Japan) includes almost all the basic phylogenetic groups which provided the material that served as the basis of the temperate floras of the northern hemisphere. As Fedorov (1957) noted, 'the so-called Holarctic elements in the flora of south-west China are more numerous and more diverse there than anywhere else in the remaining Holarctic region' (p. 29).

The temperate flora of the Himalayas, Assam, Burma and east Asia is indeed extraordinarily diverse and distinctive, and is extremely rich not only in numerous endemic species but also in many endemic genera. Amongst the endemic woody genera are such survivors of the ancient flora as *Trochodendron*, *Tetracentron*, *Decaisnea*,

Holboellia, *Akebia*, *Sinofranchetia*, *Sargentodoxa*, *Nandina*, *Sinomenium*, *Cercidiphyllum*, *Euptelea*, *Disanthus*, *Corylopsis*, *Sinowilsonia*, *Eucommia*, *Hemiptelea*, *Ostryopsis*, *Rhoiptelea*, *Platycarya*, *Dipteronia*, *Pteroceltis*, *Broussonetia*, *Sorbaria*, *Stranvaesia*, *Davidia*, *Helwingia*, *Aucuba*, *Kalopanax*, *Clematoclethra*, *Idesia*, *Pterostyrax*, *Dipelta*, *Kolkwitzia*, *Weigela* and many others. As Li (1953, p. 216) pointed out, the area of the greatest concentration of relict woody genera extends from Szechuan in west China eastwards across the valley of the Yangtze to Japan, the region richest in the temperate flora being western Hupeh and eastern Szechuan. However, floristic richness and high endemism are only part of the story.

It is well-known that in east and south-east Asia (especially south-west China) the transition from the subtropical montane flora and vegetation to the temperate, on one hand, and the tropical, on the other, is remarkably gradual. Thus Fedorov (1957, p. 29), who studied the vegetation of south-west China for some years, is of the opinion that the forest of Yunnan is clearly transitional, being intermediate in character between tropical and Holarctic vegetation, and states, furthermore, that 'all these elements of the flora are here genetically and systematically interconnected and do not merely exist together in the same place'. In 1966 I observed in Assam very gradual and almost imperceptible transitions from the tropical to the subtropical forests. Numerous existing phylogenetic series can be observed in the flora of the eastern Himalayas, Assam and south-east Asia; there is an abundance not of isolated 'missing links', but of whole chains of such links, and this makes it possible to establish the pattern and direction of the morphological and ecological evolution of several exceedingly important lines of angiosperm development. The phylogenetic series of forms transitional between the subtropical and temperate elements, which are clearly shown in many genera, families and even orders, are especially pertinent to the problems with which we are concerned. It is the only region of the world in which we can observe so great a concentration of primitive representatives of the Holarctic temperate flora and, at the same time, so many phylogenetic links (intermediate forms) in taxonomic units of such diverse categories. A phylogenetic and phytogeographical analysis of some basic components of the flora of this region shows clearly that it is not merely a 'refugium of Tertiary relicts' but the cradle of the temperate Eurasian flora. The origin of deciduous temperate woody forms from evergreen subtropical basic types is shown here with astonishing clarity and completeness.

The presence of entire phylogenetic series linking temperate forms with subtropical and tropical ones in the eastern Himalayas, Assam, Yunnan, upper Burma, North Vietnam and eastern Asia is evidence that this part of the Asian continent is the primary centre of origin of the temperate flora of the northern hemisphere, and not simply a refugium of relicts or secondary centre of development, as many authors have thought. In refugia, especially if they are situated at any distance from the original distribution centre, we find only isolated surviving elements of phylogenetic series; this is undoubtedly a result of the operation of purely statistical processes, for it is obvious that the likelihood of any prolonged migration of whole phylogenetic series, including both the basic primitive forms and the derived types, is extremely small. The elementary units of any gradual and prolonged migration are of course populations. For the most part, individual populations are the units of dispersal. If the populations are biotically independent (as, for example, the plants of an open community), then they are able to migrate as independent units. If, on the other hand, they are distinctly interdependent biotically, e.g., the plants of the forest ground layer and shade-tolerant woodland herbs, specialised parasites, etc.) then their migration proceeds communually. Therefore, migration of whole communities occurs as well as of individual populations, but the probability that an entire flora, with all its elements and phylogenetic series, will be dispersed as a whole is exceedingly small. In fact, the migration of a member of any phylogenetic series is in most cases independent of the migration of the remaining members, and of course the probability of the coincidence of a considerable number of such independent events is very small. On the other hand the independent dispersal of the individual members of a series is much more likely. For this reason, only odd chance members of different phylogenetic lines are usually found in typical Tertiary refugia situated far outside the original distribution centre; series of taxa are not found, apart from a few youthful species series. This can be observed not only in small refugia, like the Balkan peninsula, Colchis and Hyrcania (Talysh), but also in one as large as Atlantic North America. The probability distribution of the species composition of a refugium is fully in accordance with elementary probability theory. In the flora of east and south-east Asia we observe a multitude of comparatively complete phylogenetic series, which start with series of species and finish with series of families and even orders. The most remarkable example of this type is afforded by the family *Magnoliaceae*.

CHAPTER 14

THE DIFFERENTIATION OF FLORAS AND THE DEVELOPMENT OF THE MAJOR PHYTOCHORIA

Zonation and Major Phytochoria in the Cretaceous Period

Although the origin of the angiosperms dates at least as early as the first half of the Jurassic period, their extensive dispersal took place only during the Cretaceous period. In the very first stage of their dispersal a definite geographical zonation was established in their distribution as well as certain regional differences within each zone. The former idea that the Cretaceous flora was of a 'mixed' nature, not yet differentiated into temperate and tropical genera, does not reflect our present state of knowledge. It is true that the Early Cretaceous floras are as yet so insufficiently studied, the determinations of Early Cretaceous genera so uncertain, and the fossil floras themselves so few and so poor, that it is still difficult to say anything really definite about their zonation and regional differentiation. However, the presence, for example of distinctive angiospermous forms in the Early Cretaceous deposits of New Zealand (the Paparoa strata in North Westland) along with typical southern ferns and several conifers (*Araucaria* and *Podocarpus* spp.) is evidence (Oliver, 1955) of the apparent existence at that time of differences between the extratropical floras of the northern and the southern hemispheres. According to Berry (1937) and Axelrod (1952), the temperate floras of the northern and southern hemispheres were already differentiated by the middle of the Cretaceous period. Axelrod distinguishes three important early lowland Cretaceous floras—the Tropical Cretaceous, the Arcto-Cretaceous, and the Antarcto-Cretaceous—of which the tropical is the basic one and the other two derived from it. From the Cenomanian onwards, the differentiation of floras increased, zonal and regional differences becoming more pronounced. For example, pollen grains and leaf impressions of *Nothofagus*, which now occurs exclusively in the southern hemisphere, have been found in the Late Cretaceous of New Zealand and Antarctica, whereas no authentic remains of *Fagus* have been found in the southern hemisphere (Cranwell, 1963). In the northern hemisphere authentic remains of *Nothofagus* are unknown,* but *Fagus* is often found and is known from the middle of the Cretaceous. As for the numerous reports of the existence, in Cretaceous and Tertiary deposits, of many genera characteristic of the temperate flora of the southern hemisphere, the vast majority of them are erroneous, and a considerable number of the determinations have already been discounted.

The main differences between tropical and temperate angiosperm floras were already manifest in the Late Cretaceous, and within the limit of the Cretaceous Holarctic flora two major phytogeographical regions were distinctly outlined—the Boreal and the Tethyan.†

The northern temperate and cold parts of the Late Cretaceous Holarctic were occupied by the vast Boreal-Cretaceous region.‡ A purely temperate flora was dominant in this region, consisting mostly of mesophilous broad-leaved, mainly deciduous trees and shrubs, such as species of *Acer*, *Alnus*, *Betula*, *Cocculus*, *Corylus*, *Credneria*, *Fagus*, *Grewiopsis*, *Lindera*, *Magnolia*, *Platanus*, *Quercus*, *Sassafras*, *Viburnum*, *Ziziphus*, etc., as well as *Ginkgo*, various conifers (including *Agathis borealis* and *Sequoia*) and ferns. This region occupied a vast tract of land corresponding to the territory of north and north-east Europe, Kazakhstan, Siberia, the Far East of the U.S.S.R., Japan, Korea, part of North America, and the Arctic (with the possible exception of its polar portion). In Europe the southern boundary of the Boreal-Cretaceous region was deflected

* Reports of finds of pollen grains of *Nothofagus* in the Cretaceous deposits of Siberia and the Far East of Russia are based on erroneous determinations.

† In the remote geological past subkingdoms certainly originated as regions or even as provinces. Therefore, while speaking about the Late Cretaceous and the Early Tertiary I shall call them regions.

‡ Vakhrameev (1957) called it the Siberian, and in 1961 I called it the Angaro-Cretaceous, but the name Boreal-Cretaceous is more appropriate, as it covered a wider area than Siberia and its boundaries stretched beyond the limits of the Angara continent (Takhtajan, 1966).

northwards by a few degrees, probably mainly as a result of the influence of the Palaeogulfstream. Therefore the temperate flora occupied mostly the central and arctic parts of Europe. Within the Boreal-Cretaceous area, there were already certain regional differences, for example, the Late Cretaceous fossil floras of western Siberia and western Kazakhstan were characterised by their very insignificant content of ferns and gymnosperms, whereas in the fossil floras of Primorie and Sakhalin ferns and gymnosperms were dominant, amongst which Mesozoic elements like *Cladophlebis* and *Nilssonia* played a conspicuous part (Yarmolenko, 1935; Baikovskaya, 1956; Pokrovskaya and Stelmak, 1960; Vakhrameev, 1966). However, the Late Cretaceous Boreal flora was still comparatively homogeneous, as a result of extensive migrations and of the close links that existed between Eurasia and North America, both by the North Pacific route (through Beringia) and by the North Atlantic route (across the land-bridge that included what is now Greenland and Iceland). The link through Beringia was a particularly strong one, and served as a bridge joining North America directly to eastern Asia. In the composition of the flora of the Boreal-Cretaceous region we observe representatives of both specialised angiosperm families (the majority) and certain primitive families (e.g. *Magnoliaceae*). Several writers consider that the wide distribution of certain primitive genera in the temperate zone of the northern hemisphere during the Cretaceous period makes it impossible to establish the location of the primary centre of angiosperm expansion on the basis of the present pattern of distribution; but they fail to take into account here the exceedingly important fact that up to the Cretaceous period angiosperms were completely absent from this whole area, into which they undoubtedly penetrated from the south. If we accept that the angiosperms arose before the Cretaceous period, then it is logically certain that they arose outside this area.

To the south of the Boreal-Cretaceous region lay the Tethyan-Cretaceous region characterised by a considerable proportion of evergreen trees and shrubs, this being a result of the prevalence of a subtropical, or in places a subtropical—warm-temperate transitional, climate. As well as evergreen species, there was a certain proportion of deciduous forms but, unlike those of the Boreal-

Cretaceous region, they were comparatively narrow-leaved. In places, deciduous forms made up a fairly high percentage of the total. The ferns of the subtropical region were of a more southerly type than those of the temperate flora of the Boreal-Cretaceous region, while amongst the gymnosperms there were in several places a fair number of araucariads, sequoias and even bennettitaleans. The floras of this region contained many *Lauraceae* and evergreen *Fagaceae*, and palms were also characteristic. Differences within the region were not great, but were already clearly apparent. They were expressed mainly in the specific composition and only to a lesser extent in the genera.

The Tethyan-Cretaceous subtropical flora was characteristic of the islands lying in the area corresponding to what is now southern Europe and the Caucasus. The richest known fossil flora of this type is the one from the so-called Perutz deposits in north-west Czechoslovakia (Cenomanian), but, in spite of a number of special studies that have been devoted to this flora, our knowledge of it is far from sufficient. Although many erroneous determinations have been made and need to be revised, we can still say that this flora did contain many exotic subtropical angiosperms, gymnosperms and ferns. Cenomanian fossil floras of a more or less similar type are known from Niederschöne near Freiburg in Germany, from Portugal, south Armenia (between the villages of Gnishik and Aush), and elsewhere.

In Europe the northern boundary of the Tethyan-Cretaceous region extended across central England, southern Scandinavia, the Baltic region and central Russia to the northern part of the Caspian Sea; and thence approximately through the Aral Sea, Lake Balkhash and Mongolia to north China. The subtropical zone undoubtedly extended across Central Asia, Mongolia and a considerable part of China, but the Late Cretaceous fossil floras from these countries are almost unknown. The dominant vegetation types in these regions were most probably of a xerophilous nature (Kryshtofovich, 1954). It is now apparent from lithological data, that during the Late Cretaceous a broad arid zone extended from Spain and North Africa through west and central Asia to eastern China and northern Indo-China (Strakhov, 1960); it covered the southern part of the subtropical zone and the northern part of the tropical zone. Some of the most convincing evidence for the existence of this zone is the presence of gypsum and salt in the Late Cretaceous deposits of Spain, the northern Sahara (from Dakar in the west to the shores of the Red Sea in the east), Sinai, Palestine, Syria, Iraq, Arabia, Tadzhikistan, Fergana, north and central China, Laos and Annam, and also the presence of red beds in the Gobi, in Silan and in many places in China. Thus 'we may visualise a vast latitudinally extended arid region closely corresponding to the analogous region of the Caino-zoic' (Strakhov, 1960). In the area which is now southern Europe, north Africa and western Asia there were no broad tracts of land but only islands, and in all probability conditions were not suitable for the development of xerophilous vegetation. On the other hand, the arid zone occupied an extensive area of the Angara continent, including central Asia and an appreciable part of China, especially its north-western and northern regions. These arid zones of Angara-land were probably among the most ancient centres of development of the xerophilous angiosperm flora.

Most of the evidence for the existence of vast arid regions with xerophilous vegetation in the Tethyan-Cretaceous region is indirect. This is understandable, if we take into account the fact that conditions in arid regions are very unfavourable for the burying and fossilisation of plant remains. The possibilities of fossilisation in an arid climate have been critically analysed by Kryshtofovich (1954), who wrote: 'Preservation of the usual plant remains (leaves and stems) must, of course, be an extremely rare event in the case of steppe, and especially desert, plants. Leaves of the terrestrial herbs, which wither and shrivel on the stem or fall amongst the turf, are usually not preserved. Leaves of the trees and shrubs of arid communities living at some distance from lakes and marshes fall and are destroyed on the spot, and so do not reach basins of deposition. From time to time chance burial, usually of vegetable detritus, may occur in deposits of temporary lakes or storm waters. Often such remains are destroyed during transportation by the desert wadis, and the deposits that contain them are in their turn subject to destructive erosion. The accumulation of air-permeable loess and loams is also a process which does not provide favourable conditions for fossilisation. The most favourable circumstances for the preservation of plant remains in arid regions are being buried under volcanic ash (a rare occurrence) or mud pouring out on to the intermontane lowlands, or becoming permeated by petroleum. Remains of plants of xerophytic communities may also be found in the sediments of lakes—in papery shales—and we know of such cases in the Jurassic and Cretaceous periods.' Therefore, the ordinary method of the palaeobotanist—the study of leaf-impressions—is little used here, and much more information can be obtained from the study of fossil spores, pollen grains, fruits and seeds.* But there have been few such studies yet. Therefore, in considering the xerophilous floras of the past, we still depend mainly on indirect evidence and in particular on the evidence of palaeozoology. Very important results have also been obtained from research on the floras of contemporary arid regions—study of their endemism and systematic connections and analysis of the distribution areas of their more important constituents. These investigations have definitely shown the great antiquity of the flora of the deserts of Asia, which probably arose in the Late Cretaceous epoch—see Engler (1914), Popov (1927), Grubov (1959) and Lavrenko (1962). The initial nucleus of the Late Cretaceous xerophilous flora of Asia and likewise the floras of the arid regions of Australia, Africa and America probably had an east and south-east Asian origin.

South of the subtropical Tethyan-Cretaceous region lay the vast kingdom of the Palaeotropical Cretaceous flora. The basic differences between the Holarctic and the tropical floras were already clearly apparent in the Late Cretaceous. From the Late Cretaceous (Senonian and Danian) deposits of Nigeria, Egypt, Syria, Iraq and Iran, remains of strictly tropical plants are known, indicating the existence in this region of tropical floras of the Indo-Malesian type (Chandler, 1954; Chesters, 1955; Edwards, 1955). Remains of mangrove vegetation are known also from the Late Cretaceous

* The study of plant detritus, including various 'diaspores' (disseminules or units of dispersal), such as fruits, seeds, conifer cones and the megaspores of heterosporous pteriodophytes is inappropriately known as 'palaeocarpological analysis'. It would be more accurately called 'diaspore analysis' or 'palaeodiasporology'.

deposits of Austria (Hoffman, 1948). This flora which had probably migrated from south-east Asia along the ancient Tethys Sea (Thiselton-Dyer, 1878; Reid and Chandler, 1933), flourished on its southern shores and islands.

Edwards (1955) was right in emphasising 'the immense importance of the Tethys Sea as a migration route for plants'. During its prolonged history the Tethys, both by marine currents and by migration along its maritime fringes, must, as Edwards says, 'have been an extremely important agent in the latitudinal spread of land plants'. This intensive migration of tropical and subtropical floras along the Tethys was a result of the free communication with the Indian Ocean which, unlike the present Mediterranean, it possessed. As the tropical flora spread along the southern shores of the Tethys, so the evergreen subtropical and deciduous temperate floras spread mainly along the latitudinal chains of mountains to the north of it. The bulk of the purely temperate flora, including most of the cold-tolerant forms, advanced by other routes, however, and colonised the vast areas of Angara and the Arctic. This dispersal of the temperate flora, which began in the early Cretaceous, went on for a very long time, but the routes by which migration took place are still not fully clear.

Study of the generic and family endemism of contemporary tropical floras leads to the conclusion that the differentiation of the tropical flora of the world into two kingdoms—the Palaeotropical and the Neotropical—goes back to the Late Cretaceous. Each of these two kingdoms has a series of endemic families (see Appendix II), some of them extremely ancient, and each is distinguished by a very high percentage of generic endemism. Engler (1882) came early to the conclusion that amongst the numerous genera comprising the floras of these two kingdoms the number of common genera was relatively small. According to Good (1964) the total number of pantropical genera is only about 250. This small number of common genera is evidence that effective connection between the tropics of the Old and New Worlds was broken very early.

The extratropical regions of the southern hemisphere were occupied by subtropical, warm-temperate and temperate floras; we still know very little about their composition. Hooker (1860) had already concluded that the many affinities between the three southern floras—the Antarctic, the Australian and the South African—indicated that they could have been members of one great flora which at one time may have occupied a very large area. The distributions of many genera and even families clearly indicate the existence of an ancient bond between the present floras of the extra-tropical regions of the southern hemisphere. It will be sufficient to point to the distributions of such families as the *Philesiaceae*, *Centrolepidaceae* and *Restionaceae*, and such genera as *Astelia* (*Liliaceae*), *Donatia*, *Drimys*, *Eucryphia*, *Griselinia* (*Cornaceae*), *Hebe* (*Scrophulariaceae*), *Laurelia* (*Monimiaceae*), *Leptocarpus* (*Restionaceae*), *Lomatia* (*Proteaceae*), *Nothofagus* and *Oreobolus* (*Cyperaceae*) (see van Steenis, 1962).

The existence of an ancient Antarctic centre of development for the temperate flora of the southern hemisphere is accepted by many authors—Hill, (1929), Wulff (1944) and van Steenis (1962), amongst others. There is no reason to doubt the origin of this flora from the tropical flora; its basic phylogenetic lines lead to the flora of south-east Asia, but unfortunately it is not possible to consider this topic further here. The extratropical floras of the northern and the southern hemispheres had very likely a common origin, but they were already differentiated in the Cretaceous period.

One can hardly doubt that the phytogeographical kingdoms of the southern hemisphere were already discernible in the Late Cretaceous, but the paucity of the palaeobotanical material does not allow us to present decisive evidence in favour of this. However, the floristic differences between these kingdoms are so great that their origin could hardly be more recent; their differentiation most probably began in the Cretaceous period.

The Modernisation of Angiosperm Floras and the Development of Phytochoria during the Tertiary Period

A rapid modernisation in generic composition took place amongst the angiosperms at the end of the Cretaceous period and in particular during the Palaeogene. While we find in Cretaceous fossil floras many organ-genera or even form-genera, e.g. *Credneria*, *Dalbergites*, *Cissites*, *Protophyllum*, *Menispermites*, *Nordenskioldia*, *Ptero-*

spermites, *Querexia*, etc. of more or less uncertain taxonomic position, the majority of the species of Eocene floras are referable to living genera, and many even show affinities with existing species. This is confirmed not only by the study of leaf impressions but also especially convincingly by the study of fossil fruits and seeds, i.e. by diaspore analysis. The angiosperm floras of the Oligocene consist almost entirely of existing genera; and the floras of the Neogene closely approach those of the present, even in their specific composition. In the Miocene, and particularly the Pliocene, the angiosperm floras are so highly modernised that many species are practically indistinguishable (or distinguishable only with difficulty) from modern ones. Likewise, the vegetation of the Tertiary (especially the Neogene) most closely approximates to that of the present, and this is an indication of the ecological similarity of their component species.

At the beginning of the Tertiary period the geographical differentiation of the angiosperm floras was intensified, and the phytogeographical kingdoms and regions became more clearly and sharply distinct. The history of these changes has so far been studied very insufficiently, and mainly in the northern extratropical regions which belong to the Holarctic kingdom.

During the Tertiary period new provinces continued to develop; the vast Late Cretaceous Boreal region became the Boreal-Tertiary region, and the Late Cretaceous Tethyan region the Tethyan-Tertiary region.* In the semi-arid areas of south-western North

* In the works of A. N. Kryshtofovich, who has played an extremely important role in the study of phytogeographical zonation during the Tertiary period, the reader will find a rather different nomenclature for the phytogeographical subdivisions and a rather different conception of their limits and interrelationships. For example, in one of his later works, published posthumously in 1955, he distinguishes within the Holarctic Palaeocene flora two 'provinces': a northern Greenlandian (the 'Thulean' of Seward) and a southern Gelindenian (after the classic locality of the European Palaeocene flora). According to Kryshtofovich these two Palaeocene provinces became, in the Middle Eocene, two phytogeographical regions which occupied at first much the same areas as the former provinces. As in his earlier works, he called these the Turgaian and the Poltavian. However, both his concept of the limits of the Poltavian region and his nomenclature for the two regions are open to some objection. While the boundaries of

America the Madro-Tertiary flora was developing; the pattern of its history is given by Axelrod (1958).

The Boreal-Tertiary angiosperm flora consisted of many deciduous species of such genera as *Acanthopanax*, *Acer*, *Actinidia*, *Aesculus*, *Ailanthus*, *Alangium*, *Alnus*, *Amenlachier*, *Ampelopsis*, *Aralia*, *Arctostaphylos*, *Berberis*, *Berchemia*, *Betula*, *Carpinus*, *Carya*, *Castanea*, *Cedrela*, *Celtis*, *Cercidiphyllum*, *Cercis*, *Cissus*, *Cladrastis*, *Clematis*, *Clerodendrum*, *Clethra*, *Cocculus*, *Comptonia*, *Coriaria*, *Cornus*, *Corylopsis*, *Corylus*, *Cotoneaster*, *Crataegus*, *Cyclocarya*, *Daphne*, *Diospyros*, *Eucommia*, *Euonymus*, *Fagus*, *Firmiana*, *Fothergilla*, *Frangula*, *Fraxinus*, *Gleditsia*, *Hamamelis*, *Hibiscus*, *Hovenia*, *Hydrangea*, *Juglans*, *Koelreuteria*, *Ligustrum*, *Lindera*, *Liquidambar*, *Liriodendron*, *Lonicera*, *Magnolia*, *Meliosma*, *Menispermum*, *Mespilus*, *Morus*, *Myrica*, *Nyssa*, *Ostrya*, *Ostryopsis*, *Paliurus*, *Parrotia*, *Parthenocissus*, *Phellodendron*, *Photinia*, *Physocarpus*, *Pistacia*, *Platanus*, *Platycarya*, *Populus*, *Prunus*, *Ptelea*, *Pterocarya*, *Pteroceltis*, *Pyrus*, *Quercus*, *Rhamnus*, *Rhododendron*, *Rhus*, *Ribes*, *Robinia*, *Rosa*, *Rubus*, *Salix*, *Sambucus*, *Sapindus*, *Sassafras*, *Schisandra*, *Sinomenium*, *Smilax*, *Sophora*, *Sorbaria*, *Sorbus*, *Spiraea*, *Staphylea*, *Stephanandra*,

the Turgaian are clear enough and more or less correspond to those of the Boreal region as understood by us, the Poltavian region is insufficiently clearly defined by him. According to Kryshtofovich, the vegetation of the Poltavian region developed autochthonously from that of the Gelindenian province and at first occupied basically the same region. But on his map of the early Tertiary phytogeographical provinces the Gelindenian 'province' extends from Ireland, Scotland and the Leningrad region (60° N!) to Tonkin in North Vietnam. As he wrote (Kryshtofovich, 1958a) in another work, one has to allow that the Gelindenian province 'passed quite insensibly into the flora of the zone of Tertiary tropical vegegation'. It is quite evident that the Gelindenian province and its putative successor (the Poltavian region) were made up of two historically distinct parts—the subtropical zone of the Holarctic kingdom, and the tropical Indo-Malesian flora. As regards nomenclature, we consider that for phytogeographical regions and provinces of the past, our nomenclature should be based on the same principles as are employed in naming those of the present. In other words, the nomenclature of regions and provinces and of any phytochoria (to use the term proposed by Turrill to designate any phytogeographical unit) must be formed from the names of large geographical entities (e.g. continents, islands, ocean basins) and not from the localities of fossil floras. It is expedient to keep such names as Poltavian, Gelindenian and Turgaian not for the designation of phytochoria but for the different types of fossil flora. Thus, we can say of the Palaeogene fossil flora of Kamyshin that it is a fossil flora of the Gelinden type.

Stewartia, *Styrax*, *Tetracentron*, *Tilia*, *Trochodendron*, *Ulmus*, *Vaccinium*, *Vitis*, *Weigela*, *Wisteria*, *Zelkova* and *Ziziphus*. Many of them grew almost throughout this enormous region. Some evergreen species of such genera as *Mahonia*, *Epimedium*, *Hedera*, *Ilex*, *Buxus*, *Laurocerasus*, *Pyracantha*, *Rhododendron*, various *Lauraceae*, *Viscum* and *Loranthus* also occurred in the Boreal-Tertiary deciduous forests, especially in the warm-temperate parts of the region. In northern Siberia, North America and the Arctic islands some deciduous woody angiosperms reached the polar zone, where palaeobotanical evidence shows that a cool-temperate climate of oceanic type was prevalent. Southwards the cool-temperate climate gave way to a warm-temperate one, and this in turn passed very gradually into the subtropical climate of the Tethyan region. This general zonation was clearly evident throughout the whole Tertiary period, but in comparison with today the zones were displaced more or less to the north. The boundaries of these zones were not stable, however, but were subject to appreciable fluctuation.

To the south of the Boreal-Tertiary region lay the Tethyan-Tertiary region, which was characterised by subtropical vegetation consisting mainly of evergreen trees and shrubs. The herbaceous ground cover was probably much poorer in the subtropical forests of the Tethyan region than in the forests of the temperate zone, but, on the other hand, there were rather more lianas and also some epiphytes. In most cases the leading role in these forests was played by numerous *Lauraceae* and thermophilous mostly evergreen *Fagaceae*. Characteristic of the flora of this region were such genera as *Cinnamomum*, *Litsea*, *Neolitsea*, *Persea*, *Laurus*, *Quercus* (including *Cyclobalanopsis*), *Castanopsis* and *Lithocarpus*, and also various subtropical palms, especially *Trachycarpus*, *Livistona*, *Chamaerops* and *Sabal*.

To the north and in the mountains the subtropical forests gradually passed into forests of a temperate type, while on the shores of the Tethys they contained in the Palaeogene a considerable number of tropical elements. In the Palaeocene and especially Eocene the northern limit of subtropical vegetation was advanced quite considerably (but usually by not more than 15°) to the north (Fig. 30, p. 197), while the tropical vegetation was advanced to a less extent (by not more than 10°). As the numerous palaeobotanical data show, a zone of purely tropical vegetation did not occur in Europe or (especially) in north Asia, although isolated tropical elements (or in places in Europe even the mangrove communities of tidal tropical shores) did penetrate far to the north, reaching even southern England in the Eocene. It is difficult for a botanist who has seen real tropical vegetation to accept any of the Tertiary fossil floras of Europe as tropical. Thus, when the eminent American palaeobotanist, Berry, had spent some time in tropical South America, he returned convinced that 'none of the fossil floras of the Temperate Zone that palaeobotanists have termed "tropical" are tropical in the strict sense of the word' (Berry, 1929). Barghoorn (1953), in his interesting article on climatic changes in the light of palaeobotanical evidence, came even more definitely to the conclusion that the existence of truly tropical conditions in middle latitudes during the Eocene was quite unlikely. In his opinion, it was more likely that an absence of freezing allowed an unusual extension poleward of tropical plants. 'Thus,' wrote Barghoorn, 'the situation was featured by a gentle temperature gradient between high and low latitudes instead of the steep gradient which now exists. Hence the absence of cold, the lethal factor to tropical vegetation, rather than extreme warmth is indicated' (p. 244). Many contemporary palaeobotanists and geologists have now come to similar conclusions. For example the Russian geologist Strakhov (1960, p. 168) holds that 'the general climatic zonation of the Palaeogene was the same in its basic outline as it was in the Neogene and still is today; the differences concerned only details, and are easily accounted for by differences in the outlines of oceans and continents and in the relief of the land-masses'. But this, as Strakhov says, does not in the least imply that the climate of the Cainozoic was distinguished by constancy of mean annual temperature and humidity. On the contrary, there are, as he has shown, distinct signs of marked fluctuation in both. Yet, notwithstanding all this palaeobotanical and geological evidence, the botanical myth of the existence in the Palaeogene of a tropical zone in Europe, Transcaucasia and central Asia is still perpetuated, especially in popular literature. In fact, in these regions there existed, and in many places flourished, an evergreen subtropical vegetation composed of elements of the Holarctic (northern extratropical) flora, although in places there was an admixture (sometimes appreciable) of palaeotropical elements. Such penetration of tropical plants into the subtropical zone can be seen even today for example in south China. Thus the Sun-jo-Shan forest 125 km north-east of Canton, which is essentially subtropical and dominated by species of *Castanopsis*, has also a number of tropical species such as the palm *Daemonorops margaritae* and the legume *Ormosia henryi*. There are even more tropical species in the subtropical forests of the Tin-Hu-Shan some 90 km west of Canton, where such typical tropical elements as *Pothos repens*, *Dischidia chinensis* and the palm *Caryota ochrandra* are quite conspicuous. A widespread penetration by tropical species into the subtropical forest can also be observed in Assam, Yunnan and Tonkin. Moreover, certain tropical plants (members of the family *Zingiberaceae*) can be found even in the temperate forests of east China. These facts show that it is necessary to consider the entire complement of a fossil flora, not only isolated representatives of it, in order to resolve questions concerning the general character of plant communities and the zonation of climates and vegetation in the past. When this is done, it is clear that even the exceptionally rich and varied Early Eocene London Clay flora is not fundamentally tropical, though it is characterised by a plentiful admixture of tropical elements. The penetration by tropical elements so far north is evidently explained by the powerful influence of the Palaeogene Gulf Stream.* In the Eocene the territory corresponding to the British Isles was within the 'Gulf of Winter Warmth', as it is today. Tropical elements are much fewer in the Late Palaeocene and Eocene flora of the Paris basin and in the Eocene floras of the Ukraine, and especially in the Late Palaeocene

* According to van Steenis (1962, p. 278), 'The European Eocene deposits got their truly tropical component from allochthonous drift material carried by the Tethys.' But Chandler (1964, pp. 87-89) refutes the arguments used by van Steenis. Pollen grains of *Nypa* found in the London Clay are especially important, as 'heavy pollen grains do not travel far from their place of origin'. In his paper on the genus *Nypa*, Tralau (1964, p. 24) comes to the conclusion that its autochthony in the Eocene of Europe can scarcely be doubted.

floras of the Gelinden-Kamyshin type, although one should note that phytogeographical characterisation of these floras will be possible only after their critical revision by the newer methods of palaeobotanical research. Mangroves with *Nypa* palms grew by the shores of the warm nummulitic Tethys Ocean under favourable circumstances. Such '*Nypa*' floras, which in Eurasia are met with from London to Kiev and the Odessa region, represented northern intrazonal islands of tropical mangrove formation. As Chaney (1940) pointed out, in the Eocene the tropical floras were everywhere advanced northwards by a similar distance beyond their present-day limits. There is no doubt that the major zones of vegetation, especially the tropical, were much more stable than they were thought to be by writers in the nineteenth and early twentieth centuries, the 'romantic' period in the history of palaeobotany.

In North America, corresponding to the Old World Tethyan-Tertiary region, was the Madro-Tertiary region (geoflora) as established by Chaney (1944) and Axelrod (1950, 1958); it lay between the Boreal-Tertiary region and the Neotropical kingdom. According to Axelrod, the Madro-Tertiary flora had appeared by the Middle Eocene, and probably occupied much of the southwestern United States and adjacent Mexico by the close of the Oligocene. It extended its range northward and southward, as well as eastward and westward, in response to an expanding dry climate in the succeeding Miocene epoch, and it attained an even wider distribution during the Pliocene as the semiarid climate continued to spread. The Madro-Tertiary region was characterised by a dry climate with hot summers; sclerophyllous and microphyllous elements predominated in the vegetation. The vegetative cover was formed by species similar to those that grow today in the semiarid live oak–conifer woodland, chapparal, xerophilous subtropical bushland, semidesert and desert communities of southwest North America (Axelrod, 1958, 1960).

During the Tertiary period the differences between the tropical floras of the Old and New Worlds steadily increased. The connection between the Neotropical and Palaeotropical kingdoms was broken earlier than that between the Tethyan-Tertiary and Madro-Tertiary floras. Exchange of plants of tropical origin probably occurred mostly by way of the subtropical (both northern and southern) and to a lesser extent the warm-temperate zones. Therefore, representatives of palaeotropical families in the Tertiary deposits of America are found mainly in regions which lay in the subtropical or warm-temperate zones. One could hardly expect to find fossil remains of such palaeotropical families as *Dipterocarpaceae*, *Ancistrocladaceae*, *Didiereaceae*, *Nepenthaceae* and *Salvadoraceae* in tropical America, just as there is little likelihood of finding representatives of the *Lacistemataceae*, for example, in the Tertiary deposits of the Old World tropics. Direct connections between the tropical floras of the two hemispheres were ruptured very early, though probably not completely. It is quite possible that there was still some connection between South America and Africa at the beginning of the Tertiary period, a supposition which is supported by the highly interesting disjunct distributions involving tropical America and tropical Africa, to which Engler's (1905) interesting work was devoted; see also Wulff (1943); Croizat (1960); van Steenis (1962); and Good (1964). One of the most remarkable examples of the affinity between the floras of tropical west Africa and tropical America is the distribution of the family *Bromeliaceae*. The bromeliads make up one of the most characteristic elements of the neotropical flora, and all their species occur in tropical America, with the exception of *Pitcairnia feliciana*, which is endemic in Guinée. An analogous distribution is shown by the family *Rapateaceae* (*Commelinales*); most of its species occur in tropical South America, except for the genus *Maschalocephalus* which is endemic in Liberia. No less interesting is the distribution of the family *Vochysiaceae* (*Polygalales*); the majority of its genera occur in tropical America, but the oligotypic genus *Erismadelphus* is found in the virgin forests of Cameroon and lower Guinea; it is one of the most specialised genera of the family, and this seems to indicate that its ancestors probably migrated from America. In all probability the opposite is true of the family *Velloziaceae* (*Liliales*); one of its two genera—*Vellozia* (*sensu* Hutchinson, 1959)—is found in south Arabia (Hadhramaut), tropical and south Africa, Madagascar and also in tropical South America; the other—*Barbacenia*—is found only in tropical South America. *Vellozia* is characterised by free or almost free perianth-segments, while in *Barbacenia* they are united into a tube. Hence we may conclude that the *Velloziaceae* are more likely to have migrated from the Old World to the New than vice versa. Two more examples of links between the tropical flora of America and Africa are afforded by the genera *Aptandra* (*Olacaceae*), and *Thalia* (*Marantaceae*). These and many others indicate that there were ancient land areas (probably large archipelagos) within the Atlantic Basin, which served as an effective migration route (see Axelrod, 1960). According to Axelrod (1960, p. 256), 'trans-Atlantic migration probably was largely completed by the close of the Early Cretaceous (Albian) if not earlier', but it seems more probable that it did not entirely cease but continued in an attenuated form right up to the beginning of the Tertiary period. The Tertiary connections between the Indo-Malesian and Neotropical floras were evidently even more tenuous and probably broke down earlier. The botanical evidence is in fact compatible with the geologists' conclusion that the Pacific island area may have been a giant archipelago in Cretaceous and Tertiary times (Ladd, 1960; Menard and Hamilton, 1963).

In the southern hemisphere four well-defined floras must have existed in the Late Tertiary: the Cape (South African), the Madagascan (Malgache), the Australian, and the Antarctic. Originally they had much in common, but their differentiation had already begun in the Late Cretaceous; the community of these kingdoms, however, is still clearly shown at the present day. The affinity between the floras of the Australian and Cape kingdoms is shown by the distribution of such families as *Proteaceae*, *Philesiaceae*, *Restoniaceae*, *Cunoniaceae* and *Ericaceae* amongst many others, though at the generic level this affinity is very weakly expressed (see Weimarck, 1941, and Burbidge, 1960). The affinities of the Australian flora with the flora of South America are stronger and are well expressed both at the family and generic levels (see Gardner, 1944, and Burbidge, 1960). There are also some very interesting relationships between the floras of Madagascar and Australasia, notably New Caledonia (Good, 1964, p. 262). Because these kingdoms had long lost their links with each other, however, they developed an appreciable amount of endemism during the

Tertiary period. At the same time, the botanical evidence indicates that Australia had a link with south-east Asia for a very long time, a link which still exists today, at least as far as species with effective means of dispersal are concerned (Diels, 1936; see also Burbidge, 1960). Some of the most characteristic genera of the Australian flora had a south-east Asian origin. Thus in the opinion of Diels (1936, p. 193), the genus *Eucalyptus* appears to be a progressive derivative of the Malesian *Syzygium* type.

CHAPTER 15

THE BASIC STEPS IN THE EVOLUTION OF THE TERTIARY FLORA OF THE EXTRA-TROPICAL REGIONS OF THE NORTHERN HEMISPHERE

Lithological, palaeozoological and palaeobotanical data together indicate that there was a temporary drop in temperature at the very beginning of the Cainozoic (e.g. Strakhov 1960, p. 169). The Palaeocene Arctic flora was therefore rather poor and consisted of deciduous dicots, *Ginkgo*, conifers and a few ferns and horsetails. As for the subtropical, and still more the tropical, elements of the Arctic fossil floras which were recorded by Heer and earlier authors, they were all incorrect determinations. As Berry (1930b) pointed out, the Tertiary Arctic floras indicate a cool-temperate climate and are devoid of any subtropical elements.

The Palaeocene flora of the Arctic region still requires further study, but the general outlines are already clear. According to Kryshtofovich (1955), 'a characteristic feature of the Palaeocene flora of the northern zone was the huge size of the leaves of certain dicots, up to 40 cm long, broad in proportion, and of a delicate texture'. Such leaves give an indication of the high humidity of the purely oceanic climate of that time. The characteristic element of the Arctic Palaeocene flora, already known from the Late Cretaceous, was *Trochodendroides arctica*, which was probably a member of the family *Cercidiphyllaceae* and perhaps, as Brown (1939) supposed, of the genus *Cercidiphyllum*. It was a very widespread plant, though there is doubt to whether it was the community dominant. A very characteristic fern was *Onoclea*, which is now a monotypic genus occurring in east Asia and eastern North America. *Ginkgo* and the deciduous conifers *Metasequoia*, *Taxodium* and *Glyptostrobus* were very widespread. Species of *Platanus*, *Quercus*, *Alnus*, *Corylus*, *Betula*,

Vitaceae, *Tiliaceae*, *Juglandaceae* and the problematical organ-genus *Macclintockia* were also characteristic of this zone. Arctic Palaeocene fossil floras are known from Greenland, Ireland, northern Asia and North America.

South of the Arctic zone lay a zone of warm-temperate Palaeocene vegetation which apparently occupied a large area of Siberia, north-east Asia and North America. Further south it passed into purely subtropical vegetation such as is so richly represented in Europe. In western Europe, however, the warm-temperate vegetation zone narrowed, and in this region, particularly in England, the subtropical floras came almost to border on to the cool-temperate.

The northern boundary of the subtropical flora of the Palaeocene Tethyan region ran through Europe from southern England and Belgium to the Baltic republics and the southern Urals; farther east it evidently continued across southern Kazakhstan, north China and Japan, but we have no direct evidence of it as yet. Thus in Western Europe the northern boundary of the subtropical flora was shifted some distance to the north, and this is explained by the influence of tropical currents in the Tethys.

The westernmost locality for fossil remains of the Palaeocene subtropical flora is southern England. Study of the plant remains from deposits of Sparnacian age (Reading Beds and Woolwich Beds) led Chandler (1961) to conclude that the climate had been very warm and wet, but subtropical rather than tropical (as Gardner (1878) had earlier indicated). The only definitely tropical angiosperms among the plant remains of the Sparnacian deposits of England are *Oncoba variabilis* (Bowerb.) E. M. Reid & Chandl. (*Flacourtiaceae*), close to the living species *O. spinosa* Forssk., which grows in tropical Africa and Arabia (Chandler, 1961) and *Mastixia* (Chandler, 1964). As for *Natsiatum eocenicum* Chandl. (*Icacinaceae*), which is also considered by Chandler to be a tropical plant, it should be referred to *Hosiea*, a subtropical genus which is very close to the tropical genus *Natsiatum* (Takhtajan, 1966). There are only two living species of *Hosiea*—*H. sinense* (Oliv.) Hemsl. & Wilson in central China (Szechuan, W. Hupeh & Hunan), and *H. japonica* (Makino) Makino in the mountains of Japan (Honshu, Shikoku and

Kyushu). According to Chandler, the Chinese *H. sinensis* (which she cites under the name of *Natsiatum sinense* Oliv.) is closest to the fossil species. Thus the systematic affinities of *Hosiea eocenica* (Chandl.) Takht. (*Natsiatum eocenicum* Chandl.) quite definitely indicate a subtropical origin. Most of the other species mentioned by Chandler also have extratropical affinities (subtropical and partly temperate), e.g. *Sequoia couttsiae* Heer, *Libocedrus adpressa* J. Gardn., *Carpinus davisii* Chandl., *Liquidambar palaeocenica* Chandl., *Trochodendroides smilacifolia* (Newb.) Krysht. (cited by Chandler as '*Hamamelidaceae* ? Genus ?'), *Phellodendron costatum* Chandl., *Vitis pygmaea* Chandl., *Abelia palaeocenica* Chandl., and *Sambucus* sp.

The Palaeocene flora of north-east France was also subtropical, with an admixture of certain tropical elements. The predominant families were *Fagaceae* and *Lauraceae*, represented by various genera, both evergreen and deciduous. There were also various *Aceraceae*, *Alangiaceae*, *Araliaceae*, *Arecaceae*, *Betulaceae*, *Caprifoliaceae*, *Celastraceae*, *Cornaceae*, *Elaeocarpaceae*, *Juglandaceae*, *Magnoliaceae*, *Moraceae*, *Myricaceae*, *Myrtaceae*, *Pandanaceae*, *Poaceae*, *Rhamnaceae*, *Salicaceae*, *Sterculiaceae*, *Symplocaceae*, *Tiliaceae* and *Ulmaceae* amongst others. The conifers were represented by various *Taxodiaceae* and *Cupressaceae*, and the ferns by species of *Adiantum*, *Anemia*, *Asplenium*, *Blechnum*, *Cyathea* and *Lygodium*. The flora on the whole was of a subtropical type.

One of the most interesting Palaeocene fossil floras of western Europe is the flora of Gelinden in Belgium (Saporta et Marion, 1873, 1878) which is of Early Thanetian age. As in the Palaeocene floras of north-east France, the principal families are here *Fagaceae* and *Lauraceae* and this emphasises the subtropical character of the flora. The *Fagaceae* are represented by several evergreen and deciduous species of *Quercus* and some species of the organ-genus *Dryophyllum*. The *Lauraceae* consist of *Cinnamomum*, *Apollonias tetrantheroidea* (Sap.) Takht., *Persea palaeomorpha* Saporta & Marion, *Ocotea apicifolia* (Saporta & Marion) Takht. (=*Oreodaphne apicifolia* Saporta & Marion), *Litsea* spp., *Laurus omalii* Saporta & Marion, and *Neolitsea* sp. (see Takhtajan, 1966). *Celastrophyllum crepinii* Saporta & Marion with its characteristic coriaceous leaves and in particular *Dewalquea gelindenensis* Saporta & Marion, a problematical organ-genus referable in all probability to the *Araliaceae*, are also very frequently encountered.

The Russian Palaeocene fossil floras are very close to the flora of Gelinden and have some common or closely related species and similar ecological features. One of the richest is the flora of Ushi mountain near Kamyshin in Povolzhie, which still remains very little studied in spite of the many investigations of Palibin, Krassnov and Baranov. As in the Gelinden flora, *Fagaceae* and *Lauraceae* have a dominant role. Various species of *Quercus*, for example *Q. kamischinensis* (Goepp.) Ung., occur very frequently. Amongst the *Lauraceae* we may note *Cinnamomum* spp., *Litsea magnifica* Saporta, *Persea palaeomorpha* Saporta & Marion, and *Persea* (*Laurus*) *delessii* Saporta. Various species of *Magnolia*, *Betula gypsicola* Saporta, *Fagus deucalionis* Ung., *Ilex stenophylla* Ung., *Viburnum volgense* Krassn. and species of *Dewalquea* are also characteristic. Several rich fossil floras of the Kamyshin type are to be found also on the eastern slopes of the Urals and in the eastern districts of the Orenburg district in the basin of the River Or. The Palaeocene flora of the Southern Urals (along the River Romankul) is of the same type. Unfortunately this flora has been studied even less than the flora of the Ushi mountains.

The Palaeocene floras of Europe, from the Paris Basin and Belgium to the southern Urals, in all probability represent a distinct phytogeographic province within the Tethyan-Tertiary region. The flora of this province was similar to the present flora of certain areas of Assam and south-west China. The systematic affinities and ecological peculiarities of the flora of this province indicate that it was derived from a subtropical flora of the Assam–Upper Burma–Yunnan type.

The Eocene floras are considerably better known than the Palaeocene floras, and our picture of the phytogeography of this epoch is therefore much clearer. As in the Palaeocene, the displacement of the temperate zone towards the pole was much more marked than was the northward displacement of the tropical and subtropical zones. As before, however, Europe remained an exception; there a luxuriant subtropical flora extended as far as southern England. Heer (1860) explained the rise of temperature in early Tertiary times as being the result of an Indian Ocean current which passed into the Tethys and reached central and southern Europe. Its influence was evidently stronger in the Eocene, and the latitudinal zones were therefore advanced farther to the north than in the Palaeocene. In the words of Strakhov (1960, p. 169), 'After a certain fall in temperature in Palaeocene times a marked amelioration set in during the Eocene, which considerably extended the limits of the tropical flora and of bauxite formation.' This amelioration had already begun by the end of the Palaeocene epoch (Dorf, 1960, p. 345).

The palaeogeographical conditions of the Eocene were evidently the most favourable to the development of a comparatively warm climate in the polar regions (Kerner-Marilaun, 1930; Brooks, 1950). The poles were free of ice (Schwarzbach, 1950), and woody vegetation extended almost to the poles. As Schwarzbach indicates, the polewards displacement of all the climatic zones is clearly seen on climatic maps of Early Tertiary times. The boundaries of the warmer zones were shifted polewards by 10° to 15° on the average in the northern hemisphere, and by 10° in the southern; the limits of woody vegetation were shifted even farther polewards—about 20° to 30° in the northern hemisphere and at least 10° in the southern. He therefore concludes that the temperature gradient between the poles and the equator was less than it is now, and so the climate of the world was generally more equable. The northern limit of the warmer regions was shifted northwards, most of all in Europe and western North America. In the Eocene Europe was warmed on three sides—on the south by the Tethys Sea, on the west by the Palaeogulfstream, and on the east by the 'Ob Sea', which separated Europe from Asia. The Pacific coast of North America was influenced by the former Kuro Siwo (Schwarzbach, 1950), and this explains the presence of palms in the fossil floras of Matanuska and Kupreanof Island. According to Brooks (1950), comparatively insignificant changes in the distribution of land and sea, such as often occurred during geological time, are quite sufficient to explain the differences between a 'glacial' and a 'non-glacial' climate, or between a warm geological period and a cold one. The nearer one approaches the poles, the greater is the ameliorating influence of a reduction in the area of the land. The mean annual temperature is raised mainly by a rise in winter temperatures. A reduction of 10% in land area north of 60°N would raise a 'non-glacial' January temperature by at least 3.9°C, sufficient to bring the temperature above freezing point and so introduce a 'non-glacial' climate (Brooks, 1950, p. 156). The Eocene floras of the Arctic region are much better known than the Palaeocene. The Eocene locality closest to the pole is in Grinell Land, at 81° 20′ N. There the flora is very poor, numbering, according to Heer (1878), 30 species, and according to Berry (1930b) only 15. There are no evergreen angiosperms. The floras of Spitzbergen and of King Karl's Land (east of Spitzbergen) are more southerly and a little richer, though they too show clear signs of a cool-temperate or even cold climate (Berry 1930b; Schloemer-Jäger, 1958; Manum, 1962). Among the angiosperms of the supposedly Eocene deposits of Spitzbergen we may note *Cercidiphyllum crenatum* (Ung.) R. W. Brown (found also in the Palaeocene deposits), *Hamamelis*, *Tilia*, *Alnus*, *Corylus*, *Juglans* and *Ulmus*. In spite of its more southerly situation, the supposedly Eocene (or younger?) flora of Iceland is also not distinguished by its richness. Conifers are dominant, together with *Salix*, *Alnus* and *Betula*, but *Ginkgo*, *Juglans*, *Carya*, *Platanus*, *Liriodendron*, *Acer* and *Fraxinus* (Gardner, 1885) are also met with. The rich Eocene flora of the west coast of Greenland and of Disko and Hare Islands, which has been studied by many workers, is also completely cool-temperate in character. According to Berry (1930b), *Salix*, *Betula*, *Populus* and *Corylus* are dominant, but *Ginkgo*, *Metasequoia*, *Sequoia*, *Liquidambar*, *Ulmus*, *Platanus*, *Sassafras*, *Fraxinus*, *Cornus*, *Liriodendron*, *Acer*, *Vitis*, *Alnus*, *Fagus*, *Quercus*, *Comptonia* and *Magnolia* are also found. A similar flora, also evidently of Eocene age, is known from the east coast of Greenland (Scoresby Sound, Sabin Island). The Arctic floras of North America, found at the mouth of the Mackenzie River and on the islands of the North American archipelago—Banks, Bathurst and Ellesmere, are of a similar kind.

The cool-temperate zone of the Boreal Tertiary region gradually passed southwards into the warm-temperate zone which occupied a large area of Asia and North America. Unfortunately, fossil flora localities in this zone are few, often poorly dated and mostly insufficiently studied. The phytogeographical characterisation of this zone is therefore a thing of the future.

For our purposes, however, interest lies less in the warm-temperate zone itself than in its boundary with the subtropical zone. On the basis of palaeobotanical, palaeozoological and lithological data this boundary may be traced as follows: in North America it curved southwards in a prominent arc between Kupreanof Island in the west and Nova Scotia in the east; in Europe it ran through Ireland, central England and the southern Baltic, and then ran somewhat south of the 56°N parallel of latitude to about 55°E. Thus the subtropical zone extended farther to the north than at any other time and place during the Cainozoic era. The boundary then turned south-east and ran somewhere across the Chelyabinsk district, north Kazakhstan, north China, Korea and Japan (Chaney, 1940; Kryshtofovich, 1955; Axelrod, 1960; Dorf, 1960; Strakhov, 1960) (Fig. 30).

The study of the subtropical floras of the Eocene epoch is of exceptional interest and holds the key to the solution of many phytogeographical problems connected with the development of the extra-tropical floras and vegetation of the northern hemisphere. Being unable in these pages to give any detailed characterisation of the subtropics of the Eocene, I will limit myself merely to some passing remarks on certain of the fossil floras.

The Early Eocene (Ypresian) Wilcox flora is of outstanding interest amongst the North American fossil floras. It is known from over 130 localities in south-east North America (Kentucky, Tennessee, Alabama, Mississippi, Arkansas, Louisiana and Texas). According to Berry (1937), the largest families in the Wilcox flora are the *Lauraceae*, *Fabaceae* (*Mimosoideae*, *Caesalpinioideae*, *Faboideae*), *Moraceae*, *Rhamnaceae*, *Sapindaceae*, *Sapotaceae*, *Anacardiaceae*, *Myrtaceae*, *Combretaceae*, *Juglandaceae*, *Sterculiaceae*, *Araliaceae*, *Apocynaceae*, *Celastraceae*, *Polypodiaceae s.l.*, *Arecaceae*, *Rutaceae*, and *Meliaceae*. Berry (1916, 1930a, 1937) denied the presence of strictly temperate elements in this flora, but later studies revealed the presence of the genera *Betula*, *Comptonia*, *Fagus*, *Sassafras* and *Staphylea* (Brown, 1944, 1946). The Wilcox flora is a luxuriant

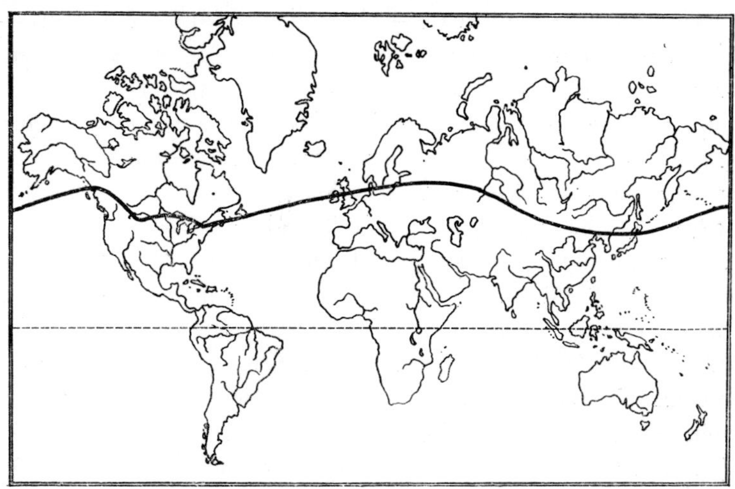

Fig. 30. The boundary between temperate and subtropical vegetation in the northern hemisphere in the Early Tertiary (mainly Eocene). (Based on Chaney, 1940, Kryshtofovich, 1954, Axelrod, 1960, Dorf, 1960, and other sources.)

subtropical flora with a significant admixture of tropical elements; 82 (60%) of the genera are represented in the present flora of the south-east United States, 72 (53%) in the flora of central and eastern China, and 93 (68%) in the contemporary flora of eastern Mexico (Sharp, 1951).

The exceedingly rich London Clay flora, numbering about 500 species (Chandler, 1961), is of the same age as the Wilcox flora. In addition to *Selaginella*, ferns and gymnosperms (amongst them *Cephalotaxus*, *Sequoia* and *Taxodium*), it contains representatives of nearly 50 angiosperm families, among which the *Anacardiaceae*, *Annonaceae*, *Arecaceae*, *Burseraceae*, *Cornaceae*, *Icacinaceae*, *Euphorbiaceae*, *Juglandaceae*, *Lauraceae*, *Magnoliaceae*, *Menispermaceae*, *Rutaceae* and *Sapindaceae* play a leading role. The *Lauraceae* are the most numerous. Such an abundance of *Lauraceae* makes it seem likely that *Fagaceae* were also abundant, but only pollen grains of *Fagaceae* and some oak wood have been found. The scarcity of *Fagaceae* in the London Clay flora is not strange, if we recall that unlike the floras of the Wilcox group it is a diaspore flora, not a leaf-flora, and is made up of the remnants of fruits and seeds—i.e. of diaspores—and not of leaf impressions. The *Fagaceae* are in an unfavourable position in this respect, as the fruits and even the cupules of this family fossilise very badly, for they decompose quickly. Besides the many purely tropical elements, there are numerous subtropical and even temperate types in the London Clay flora, such as *Magnolia* (15 species), the organ genera *Wardenia* and *Palaeosinomenium* (which are very near to *Sinomenium* and *Menispermum*), *Trochodendron* ? *pauciseminum* E. M. Reid & Chandl., *Cercidiphyllum spenceri* Brett (wood), *Corylopsis*, *Morus*, *Nyssa*, *Vitis*, *Hosiea eocenica* (Chandl.) Takht., *Vitis*, *Ampelopsis*, *Parthenocissus* and *Ardisia*. Pollen grains of *Taxodium*, *Metasequoia* and many typical temperate to subtropical genera, such as *Platanus*, *Alnus*, *Betula*, *Corylus*, *Carpinus*, *Pterocarya*, *Tilia*, *Pistacia* and *Sambucus* have also been found (Chandler, 1964). Many genera are represented by species which are close to present-day subtropical forms, for example *Cinnamomum*, *Alangium*, *Meliosma*, *Tetrastigma*, *Symplocos*, *Ehretia* and various *Juglandaceae*. On the other hand, the characteristic palaeotropical family *Dipterocarpaceae* is apparently absent. From these facts I have concluded that the London Clay flora was subtropical with a considerable admixture of tropical elements, as, for example, are the forests of Sun-jo-Shan, 125 km north-east of Canton (Takhtajan, 1961). Van Steenis (1962) came to a similar conclusion; in his opinion, the species of almost all the genera found in the London Clay could have grown in the warmer parts of the subtropics. He considers that the subtropical elements predominate over the tropical, the number of which is not great (he instances *Nypa*, *Iodes*, and, but not altogether justifiably,* the genus *Tetrastigma*). The seeds of *Rhizophoraceae* and the fruits of *Nypa burtinii* (Brongn.) Ettingsh. are of especial interest with respect to the character of the London Clay flora; they indicate the existence of mangrove vegetation in the south of England during the Eocene. Fruits of *Nypa burtinii* are also known from the Eocene deposits of France, Belgium, Italy, Poland (south of Cracow), Hungary and the Ukraine (in the Kiev area, and in Voronovka and Kalinovka in the Odessa district)—see Chandler (1964) and Tralau

* The genus *Tetrastigma* (*Vitaceae*) although on the whole tropical, has a few species which grow in the subtropics of the eastern Himalayas, Assam and east Asia.

(1964). But, as we have said before, the growth of mangroves and other tropical plants in Europe does not necessarily indicate in itself that the tropical zone extended so far north; it is rather a local effect of the influence of warm ocean currents.

The floras found in the Late Eocene Baltic ambers are exceptionally interesting and important, especially those of the Kaliningrad peninsula. Unlike the leaf flora of the Wilcox and the diaspore flora of the London Clay, its plant remains are preserved in a very distinctive manner. Parts of plants—flowers, leaf fragments, pieces of twig, etc.—which became included in a fossil resin (amber) are represented by empty casts which exactly reproduce their outlines. This form of preservation differs from that of leaves and diaspores in being less selective. In other words, the composition of this flora is more properly reflected in amber than is the case with leaves or diaspores, where fossilisation depends upon the consistency and chemical composition of the objects fossilised. Any part of the plant may be preserved in amber; even the most delicate flowers and the tiniest insects are preserved intact. Although the floras of the Baltic Eocene ambers have not been sufficiently studied, we can nevertheless establish their general character—see Goeppert and Menge (1883), Conwentz (1886, 1890) Caspary and Klebs (1907), Kirchheimer (1937), Czeczott (1961). In the amber we find mosses, ferns, various conifers and representatives of the angiosperm families *Aceraceae* (several *Acer* spp.), *Apiaceae*, *Apocynaceae*, *Aquifoliaceae*, *Araceae*, *Arecaceae* (including *Phoenix*), *Campanulaceae*, *Caprifoliaceae*, *Celastraceae*, *Cistaceae*, *Clethraceae*, *Dilleniaceae* (*Hibbertia*), *Ericaceae*, *Euphorbiaceae* (*Antidesma*), *Fabaceae s.l.*, *Fagaceae* (numerous *Quercus* spp. and species of *Castenea* and *Fagus*), *Geraniaceae*, *Hamamelidaceae*, *Hydrangeaceae* (*Deutzia*), *Iteaceae*, *Lauraceae* (species of *Cinnamomum* and the organ-genus *Trianthera* near to the living genus *Eusideroxylon*), *Linaceae*, *Loranthaceae*, *Magnoliaceae*, *Myricaceae*, *Myrsinaceae*, *Olacaceae*, *Oleaceae*, *Oxalidaceae*, *Pentaphylacaceae*, *Pittosporaceae*, *Polygonaceae*, *Rhamnaceae*, *Rosaceae*, *Rubiaceae*, *Salicaceae*, *Santalaceae*, *Saxifragaceae*, *Smilacaceae*, *Theaceae*, *Thymelaeaceae*, *Ulmaceae*, *Urticaceae* and others. It is interesting to note that Forman (1964) pointed out some resemblances between the fossil *Fagus succinea*, found in the Baltic amber, and the living genus *Trigonobalanus* (North Thailand, Malay Peninsula, Borneo, Celebes).

Towards the south-east the Eocene floras tend to become xerophilous. The plant remains of the Eocene deposits of the southern Urals are of a subtropical hemixerophilous character. According to Medea Uznadze-Dgebuadze (1948, p. 164), who studied this fossil flora, the remains indicate a mean annual rainfall of 250–500 mm or less, the maximum precipitation being in winter. She even considers that 'the forest, of which we have only a few traces, must have grown on, or not far from, the margins of a desert, on a poor sandy soil.' Among the monocots a partial remnant of a palm petiole was found. The dicots are represented by evergreen species, with coriaceous, mostly narrow, small and marginally revolute leaves. First place in number of impressions is taken by *Leucothoe protogaea* (Ung.) Schimp., which forms almost 30% of the total flora. Narrow-leaved evergreen oaks like *Quercus apocynophyllum* Heer and *Q. elaenae* Ung. are characteristic, as are species of *Cinnamomum* and *Myrica*. The Eocene floras of Badkhyz in Turkmenia also are hemixerophilous in character, especially those from the vicinity of Lake Er-Oilan-Duz, and are characterised by various *Lauraceae*, *Rhus turcomanica* (Krysht.) Korov., *Ziziphus* spp., *Diospyros*, an enigmatic form-genus *Palibinia*, *Araucarites sternbergii* Goepp., *Chamaecyparis* sp. and *Woodwardia roessneriana* (Ung.) Heer. (Vasilevskaya, 1957). The Eocene flora of the Pavlodar Priirtyshie (Kazakhstan) was less xerophilous in character. But even here the narrowness and coriaceous texture of the leaves is evidence of a dryness, at least for a certain period of the year (Budantsev, 1957, p. 178). Fossil floras with *Palibinia* are also known from Shensi and Hunan provinces of China (Tao, 1965; and Li, 1965).

The events that followed in the Oligocene epoch played a most important part in floristic and faunistic development. The distribution of land and sea became much more like what it is now, the link between the Tethys and the Indian Ocean was broken, and a general elevation began, with the formation in the Mediterranean geosynclinal region of the Alpine fold-structures. As a result, the climate of the middle and higher latitudes became increasingly colder during the Oligocene, and this, together with the other physico-geographical changes, greatly influenced the flora and vegetation and led to their appreciable modernisation. During the Oligocene, the tropical elements almost completely disappeared from Europe; a few relict forms remained only in its southern parts, in the Caucasus and in the south of central Asia, where some have persisted to the present day. At the same time, the subtropical flora gradually gave way to the temperate.

With the disappearance of the Tethys and the spread of the arid regions of Asia, the main migration route from east and south-east Asia to Europe and the Mediterranean disappeared, and long-distance migrations of the temperate flora were essentially brought to a close. At the same time the temperate flora gradually and steadily (except for a few local fluctuations and recessions) extended its area at the expense of the subtropical flora, and the northern boundary of the tropical zone approached more closely to its present position.

This comparatively rapid extension of the temperate flora gives the impression that an 'Arcto-Tertiary' flora had formed somewhere to the north and forced out a subtropical and tropical flora which retreated southwards. In fact, however, the temperate flora had penetrated these regions much earlier, during the time of the 'great transmigration' of the angiosperms, and had existed, developed and diversified (in the bosom of the subtropical flora) on the mountains of the corresponding subtropical latitudes. It is not surprising, therefore, that as early as the Cretaceous and the Palaeogene we find typical temperate forms among purely subtropical floras. It was here, in the mountains of the subtropical zone, that the temperate flora uninterruptedly evolved and formed secondary centres of development, but, growing in the mountains comparatively far from areas of accumulation, it found conditions less favourable for fossilisation than did the subtropical flora. Everywhere the temperate flora underwent evolution and specialisation and then became enriched by new elements, sometimes even assimilating new material from the subtropical and tropical floras; but the basic primary elements forming its nucleus stemmed historically from east and south-east Asia.

From the Oligocene onwards the differentiation of the extratropical phytogeographical regions of the Tertiary period proceeded much more rapidly, and in the Miocene the formation from them of the existing regions began. In the early Oligocene, with its warm moist climate, the subtropical flora was still widespread and occupied a considerable area of western Europe, the southern part of the Russian platform, central Asia and North America; but by the middle of the Oligocene the subtropical flora had begun to give way to the deciduous, a process which started in the north and gradually extended southwards. Thus in the Middle Oligocene deciduous dicots clearly predominated over evergreens in the Rhineland and in many other places in Europe. The vegetation of the Baltic was also becoming more temperate; it had never in fact been fully evergreen, but in the Eocene it evidently had a mixed character, with a predominance of broad-leaved deciduous genera and a fair content of evergreen thermophilous elements (Pokrovskaya and Zauer, 1960a, 1960b). Palynological analysis, however, indicates that in the Oligocene the vegetation of the Baltic consisted of broad-leaved or mixed coniferous—broad-leaved forest with only a very insignificant sprinkling of evergreen elements (Pokrovskaya, 1956), and this is confirmed by Budantsev and Sveshnikova (1964), who studied the Oligocene brown-coal leaf-flora of the Kaliningrad peninsula (Samland). This fossil flora is characterised by an obvious prevalence of the deciduous trees (especially by *Populus*, *Alnus*, *Nyssa* and *Rhamnus*). The same changes took place in other places, e.g., the west Siberian lowlands, where the proportion of evergreen angiosperms in the broad-leaved and coniferous forests suffered a marked reduction by the Late Oligocene. The northern boundary of the subtropical zone was constantly being displaced southwards; in the western Kazakhstan, for example, temperate vegetation had displaced the subtropical by the middle of the Oligocene (Zhilin, 1968).

In Early Miocene times the temperate flora occupied a vast area in Eurasia and North America, and its southern boundary began rapidly to approach its present position. Central and eastern Europe were occupied by broad-leaved forest, although there was still a fair admixture of evergreen forms, especially in the more southerly and coastal regions. But evergreen forms were already absent in the Late Miocene floras of western Siberia and Kazakhstan.

In the subtropical regions of Asia and in places in North America the evergreen subtropical flora continued to develop and has persisted, with some specific alteration, to the present day; it has also persisted in the Mediterranean area, though in a much more impoverished form. The Tertiary laurel-leaved forest of Europe is well known to be best preserved in the Canary Islands, where members of the *Lauraceae* form beautiful groves on the islands of Gomera, Ferro, Teneriffe and Palma (Cifferi, 1962).

There the lauraceous species grow at 500–1200 m on mountain slopes exposed to the south-east trade winds, and also in deep moist ravines (Burchard, 1929). In such forests grow *Laurus azorica* (the most important forest-forming species in the Canary Isles, which also occurs in Madeira and the Azores), *Persea indica* (the most moisture-loving of the Canary laurels), *Apollonias barbusana* (which is found mainly as an accessory constituent of the groves of Canary laurel) and, finally, *Ocotea foetens*, which flourishes in practically all the laurel forests of the Canary Islands. In the Late Tertiary all these species were widespread in the Mediterranean area, while *Persea indica* and *Apollonias barbusana* were similarly established in the Neogene (possibly Late Miocene) flora of the Goderdzi Pass in south-west Georgia. A whole series of relicts of the Tertiary subtropical flora of southern Europe and North Africa is still preserved in Macaronesia. 'The flora of Macaronesia,' wrote Wulff (1944), 'is an ancient remnant of the Tertiary flora, which has been preserved to the present day and affords us a graphic example of what the flora of the Tethyan region was like in the first half of the Tertiary period.' One must bear in mind, however, that many of the common plants of the Tethyan subtropical flora, e.g. the then very widespread genus *Cinnamomum*, have not persisted either in Macaronesia or the Mediterranean region.

One of the most important consequences of the geographical changes of the Miocene epoch was the formation of the Mediterranean flora. Hemixerophilous and sclerophyllous plant communities had already existed in the Palaeogene. For example, according to Kryshtofovich (1954), 'our Eocene flora of the Ukraine was distinguished by the toughness and narrowness of its leaves and their disorderly distribution in the sands, often in a convoluted condition'. A flora of similar type existed in the Eocene in the south of France. The Eocene floras of the southern Urals, Kazakhstan and especially Badkhyz in Turkmenia were distinctly more xerophilous, particularly the flora of Er-Oilan-Duz. Hemixerophilous floras are also known from the Oligocene, for example in Hungary, whence the late Hungarian palaeobotanist Andreánszky reported the existence of an Early Oligocene flora with very xeromorphic leaf-types. These floras, however, show a very weak genetic affinity with the present Mediterranean flora, and it is probable that only very few of their elements have entered into its composition. Elements of the Mediterranean flora began to appear only in the Miocene, and in the late Miocene the flora as a whole, with its characteristic taxonomic and ecological features, began to take shape.

The Mediterranean flora was founded mainly on the rich and varied flora of the Tethyan-Tertiary region, but elements of the Boreal flora also played a quite appreciable part in its development, especially in the east. It was built up mainly of xerophilous derivatives of subtropical and temperate elements, but initially had also a number of tropical forms, some of which have persisted as relics to the present day. Much is still obscure concerning the lines of development of the Mediterranean flora, but there is no doubt its nucleus arose autochthonously from within the Tethyan flora.

Not only the Mediterranean but also the remaining contemporary phytogeographical regions of the world (Fig. 32, opp. p. 253) developed essentially autochthonously out of those of Tertiary times. Migration no longer played the exclusively important part that it did in the formation of the botanical regions of the Cretaceous and Tertiary periods. The obliteration of the Tethys and the formation of vast desert barriers put an end to large-scale continental migration, and the spread of plants became predominantly narrowly regional in character. Similar events took place in the other continents. As a result, our floras underwent rapid differentiation and were greatly enriched by locally developed endemic groups, many of which no longer showed their systematic affinity with their ancestral forms, the immigrants from the primary centre of angiosperm dispersal.

APPENDIX I

AN OUTLINE OF THE CLASSIFICATION OF FLOWERING PLANTS

Many orders and families of flowering plants have been mentioned in this book. In different systems of classification they have often been circumscribed very differently and the systematic positions they have been accorded are likewise various. It therefore seems useful to add an appendix to this book, giving an outline of the classification that has been adopted by the author. This classification is explained in detail in my book *A System and Phylogeny of the Flowering Plants*, which was published in Russian in 1967 (in Leningrad) and to which the reader is referred. A bibliography will also be found in the above-mentioned book. Here I give only the bare bones of the system, i.e. a list of families and taxa of higher rank, together with some very brief remarks on the relationships of the orders. I have now introduced some changes both on ordinal and family levels, but the general scheme remains unaltered.

Division: **MAGNOLIOPHYTA** or **ANGIOSPERMAE** (flowering plants or angiosperms)

Class: MAGNOLIATAE or DICOTYLEDONES (dicots)

Subclass A: MAGNOLIIDAE

Superorder I: MAGNOLIANAE

Order 1: **Magnoliales (Annonales)**. The most primitive living order of flowering plants

Magnoliaceae
Degeneriaceae
Himantandraceae

Eupomatiaceae
Annonaceae
Canellaceae
Myristicaceae
Winteraceae

Order 2: **Laurales.** Near to the **Magnoliales**, but more advanced; evidently derived from some ancient vesselless representatives of the **Magnoliales**

Austrobaileyaceae
Amborellaceae
Trimeniaceae
Monimiaceae (including *Atherospermataceae*)
Gomortegaceae
Hernandiaceae
Chloranthaceae (evidently nearest to the *Austrobaileyaceae* and *Trimeniaceae*)
Lactoridaceae (occupies a rather isolated position in the order **Laurales**; this family apparently stands nearest to the *Chloranthaceae*, with which it probably had a common origin)
Calycanthaceae (nearest to the *Monimiaceae*, especially to its primitive members)
Lauraceae (probably derived directly from primitive members of the *Monimiaceae* of the *Hortonia*-type)
Gyrocarpaceae (very near to the *Lauraceae*)

Order 3: **Piperales.** Nearest to the **Laurales**. It possibly had a common origin with the *Chloranthaceae* and *Lactoridaceae*

Saururaceae
Piperaceae (including *Peperomiaceae*)

Order 4: **Aristolochiales**. Probably derived directly from the **Magnoliales**

Aristolochiaceae

Order 5: **Rafflesiales**. Probably derived from the ancestors of the **Aristolochiales**, but evidence of affinity not strong

Rafflesiaceae

Hydnoraceae

Order 6: **Nymphaeales**. Probably derived from some ancient vesselless stock of the order **Magnoliales**

Cabombaceae

Nymphaeaceae (including *Euryalaceae* and *Nupharaceae*)

Barclayaceae (near to the *Nymphaeaceae*, but appreciably more advanced)

Ceratophyllaceae (evidently related to the *Cabombaceae* and probably had a common origin with them)

Subclass B: RANUNCULIDAE

Superorder II: RANUNCULANAE

Order 7: **Illiciales** (**Schisandrales**). Evidently derived from the **Magnoliales**, most probably from the ancestors of the *Winteraceae*

Illiciaceae

Schisandraceae

Order 8: **Nelumbonales**. Usually placed in or next to the *Nymphaeaceae*, but differs in many important features, including the tricolpate pollen grains, the structure of the embryo, the presence of special respiratory openings in the pericarp and seed coat, the absence of laticiferous tubes and idioblasts, and the morphology of the chromosomes. Any direct phylogenetic link with the **Nymphaeales** is unlikely; a link with the **Illiciales-Ranunculales** stock is much more probable

Nelumbonaceae

Order 9: **Ranunculales.** Evidently derived from the ancestors of the **Illiciales**. The most primitive families of the order **Ranunculales** exhibit definite links with the **Illiciales**. The families *Lardizabalaceae*, *Sargentodoxaceae* and *Menispermaceae* are especially close to the **Illiciales**

Lardizabalaceae

Sargentodoxaceae

Menispermaceae

Ranunculaceae (including *Kingdoniaceae*?)

Glaucidiaceae (contains only one monotypic genus, *Glaucidium*, which is usually placed in the family *Ranunculaceae*; but it differs markedly from the *Ranunculaceae* in the structure of the gynoecium and ovule, in the mode of the dehiscence of the follicles, and in its caryology)

Hydrastidaceae (one monotypic genus *Hydrastis*; in many respects it occupies an intermediate position between the *Ranunculaceae* and the *Podophyllaceae*)

Circaeasteraceae

Podophyllaceae (*Podophyllum* and *Diphylleia*)

Nandinaceae

Berberidaceae (including *Leonticaceae*) (the *Berberidaceae*, *Nandinaceae* and *Podophyllaceae* have a common origin with the *Menispermaceae* and *Ranunculaceae*, and especially close links with the *Hydrastidaceae*. As well as their morphological similarity, they have in common the presence of berberin)

Order 10: **Papaverales.** Very near to the **Ranunculales**, especially to the families *Glaucidiaceae* and *Hydrastidaceae*, and likewise to the *Podophyllaceae* and *Berberidaceae*

Papaveraceae (including *Pteridophyllaceae*?)

Hypecoaceae (*Hypecoum*)

Fumariaceae

Order 11: **Sarraceniales**. A very specialised order, which has nevertheless some primitive features which place it near to the **Ranunculales**

Sarraceniaceae

Subclass C: HAMAMELIDIDAE

Superorder III: HAMAMELIDANAE

Order 12: **Trochodendrales**. In their basic chromosome number (19), their tri-multi-lacunar nodes and primitive wood structure, the **Trochodendrales** exhibit definite links with the **Magnoliales**, but in the structure of their pollen grains, flowers and inflorescences they approach much more closely to the **Hamamelidales**. In many respects the **Trochodendrales** occupy, as it were, an intermediate position between the **Magnoliales** and **Hamamelidales**, but in the totality of their characters they stand nearer to the latter

Trochodendraceae

Tetracentraceae

Order 13: **Cercidiphyllales**. Stands closest of all to the **Trochodendrales**, which it approaches also caryologically ($x = 19$), but it is distinguished from the **Trochodendrales** by the presence of vessels, very peculiar reduced inflorescences, extremely simplified flowers, monomerous gynoecium, winged seeds, pollen grain morphology and many other features. At the same time the order **Cercidiphyllales** stands relatively close to the **Hamamelidales**

Cercidiphyllaceae

Order 14: **Eupteleales**. The morphological peculiarities of the only member of the order, the genus *Euptelea*, indicate its marked systematic isolation.

Although in basic chromosome number (14) the genus *Euptelea* is nearer to the **Illiciales** than to *Cercidiphyllum* or *Trochodendron* where $x = 19$, or to the **Hamamelidales** where $x = 12$, reference of the **Eupteleales** to the superorder HAMAMELIDANAE seems to me as to be yet the only reasonably satisfactory solution

Eupteleaceae

Order 15: **Didymelales**. The Madagascan genus *Didymeles* has usually been placed near the *Leitneriaceae*, less often with the *Euphorbiaceae* or other families. Notwithstanding some similarity in pollen grain morphology (as noted by Erdtman), *Didymeles* stands very remote from the *Euphorbiaceae* (monomerous gynoecium!) and has much more in common with the *Leitneriaceae*. But the genus *Didymeles* is very sharply distinct from *Leitneria* in the structure of the inflorescence and male flower, in the unique type of pollen grain wall, in the position of the carpel, in the structure of the ovule, in the very peculiar encyclocytic stomatal apparatus, in the absence of secretory canals in the leaves and pith, in the scalariform perforation of the vessels, in the absence of wood parenchyma, in the primitive heretogenous rays, and in the fibres with larger and distinctly bordered pits

Didymelaceae

Order 16: **Hamamelidales**. This order in many ways serves as a connecting link between the **Trochodendrales** on the one hand and the 'amentiferous' orders, **Casuarinales**, **Urticales**, **Betulales**, **Fagales** etc., on the other. The order **Hamamelidales** is evidently derived from immediate ancestors of the **Trochodendrales** which had entomophilous flowers with apocarpous (but already cyclic) gynoecia and a double perianth

Hamamelidaceae (including *Altingiaceae*, *Disanthaceae* and *Rhodoleiaceae*)

Platanaceae

Myrothamnaceae (an isolated family within the order)

Order 17: **Eucommiales.** Some authors place the *Eucommiaceae* with the *Hamamelidaceae*, while others put them in the order **Urticales**, most often near to the *Ulmaceae*. But from both the **Hamamelidales** and the **Urticales** the family *Eucommiaceae* differs in such important features as absence of stipules, unilacunar nodes, pollen wall structure, the unitegmic ovule and the cellular endosperm; it is also distinguished by the production of gutta-percha. There is thus every reason for recognising it as forming a separate order, the **Eucommiales**, which was established in 1956 by Němejc. The order **Eucommiales** evidently had a common origin with the **Urticales** from the **Hamamelidales**

Eucommiaceae

Order 18: **Urticales.** This order is connected with the **Hamamelidales** and perhaps is derived directly from it

Ulmaceae (including *Celtidaceae*)

Moraceae

Cannabaceae

Urticaceae

Order 19: **Barbeyales.** The sole representative of the order, the genus *Barbeya*, has been placed in the **Urticales** in all systems; in some systems it is included in the family *Ulmaceae*, in others separated as a distinct family. From the *Ulmaceae*, as from the order **Urticales** as a whole, it differs markedly in the structure of the gynoecium, which consists of one carpel (in all *Urticales sensu stricto* the gynoecium is pseudomonomerous), and no less markedly in the structure of the pollen grain wall and in the calyx which enlarges in fruit. The opposite leaves of *Barbeya* and the absence of stipules are also features uncharacteristic of the **Urticales**. In *Barbeya* there are no cystoliths. The order evidently arose from the **Hamamelidales** or their immediate ancestors

Barbeyaceae

Order 20: **Casuarinales**. Evidently derived from the **Hamamelidales**

Casuarinaceae

Order 21: **Fagales**. In all probability derived from the **Hamamelidales**

Fagaceae

Order 22: **Betulales**. Probably had a common origin with the **Fagales**

Betulaceae (including *Carpinaceae* and *Corylaceae*)

Order 23: **Balanopales**. The presence of a cupule, the structure of the inflorescence, flowers and, to some extent, of the pollen grain wall, and likewise the wood anatomy point to a common origin for the **Balanopales** and **Fagales**

Balanopaceae

Order 24: **Myricales**. Has much in common on the one hand with the **Casuarinales** and **Betulales**, and on the other with the **Juglandales**

Myricaceae

Order 25: **Juglandales**. Has much in common with the

Myricales and also with the **Fagales** and **Betulales**

Rhoipteleaceae (in many respects the family *Rhoipteleaceae* approaches closely to the hypothetical intermediate between the *Hamamelidaceae* and *Juglandaceae*)

Juglandaceae

Order 26: **Leitneriales**. This order is evidently also one of the anemophilous derivations of the **Hamamelidales**. The elucidation of the affinity of the genus *Leitneria* is made very difficult by the extreme reduction of the flowers and inflorescences and the general specialisation of the whole plant. In the structure of the inflorescences it has much in common with the *Betulaceae*, but in *Leitneria* they are more simplified and modified

Leitneriaceae

Subclass D: CARYOPHYLLIDAE

Superorder IV: CARYOPHYLLANAE

Order 27: **Caryophyllales**. In all probability derived directly from the order **Ranunculales**. The family *Phytolaccaceae* in particular is linked with the **Ranunculales**, especially with the families *Menispermaceae* and *Lardizabalaceae*

Phytolaccaceae (including *Achatocarpaceae*, *Agdestidaceae*, *Barbeuiaceae*, *Petiveriaceae* and *Stegnospermaceae* and excluding (?) *Rhabdodendron*)

Gyrostemonaceae (near to the *Phytolacceceae*, but receptacle flat or convex, disc-like, anthers sessile or subsessile and seeds with fleshy perisperm; stems with normal growth, like the *Achatocarpoideae* and some of the *Phytolaccoideae* in the *Phytolaccaceae*)

Bataceae (near to the *Phytolaccaceae* and *Gyrostemonaceae*, but seeds without perisperm; the pollen morphology favours the assumption of a definite affinity with the *Gyrostemonaceae*)

Nyctaginaceae (an advanced entomophilous family derived from the *Phytolaccaceae*)

Molluginaceae (very near to the *Phytolaccaceae*, with which it is connected by way of the intermediate genus *Gisekia*)

Aizoaceae (very near the *Molluginaceae*, but somewhat more advanced)

Tetragoniaceae (very close to the *Aizoaceae*)

Cactaceae. Morphological investigations over the last few decades (and likewise chemotaxonomic data) have confirmed the indisputable affinity of the *Cactaceae* with the *Aizoaceae*, *Portulacaceae* and *Phytolaccaceae*. Together with its closest relative, the *Aizoaceae*, the family *Cactaceae* is in all probability derived from the most primitive apocarpous *Phytolaccaceae*. The subfamily *Pereskioideae*, characterised by an almost apocarpous gynoecium, serves as an intermediate link)

Portulacaceae (near to the *Aizoaceae* and *Cactaceae*)

Basellaceae (stands very close to the *Portulacaceae*)

Didiereaceae (has a common origin with the *Cactaceae*, *Portulacaceae* and *Nyctaginaceae*, and stands closest to the first two)

Halophytaceae (evidently stands close to the *Portulacaceae* and *Basellaceae*)

Hectorellaceae (very closely linked with the *Portulacaceae*, and possibly does not deserve separation as an independent family)

Caryophyllaceae (including *Illecebraceae*) (closely related to the *Portulacaceae*, with which it has a common origin)

Amaranthaceae

Chenopodiaceae (including *Dysphania*?). (Nearest to

the *Amaranthaceae*. Both families are related to the *Phytolaccaceae*)

Order 28: **Polygonales.** Near to the *Caryophyllales*, especially to the *Portulacaceae* and *Basellaceae*, but seeds without perisperm and with copious mealy endosperm in which the curved or straight embryo is more or less embedded. Probably derived from the same stock as the **Caryophyllales**

Polygonaceae

Order 29: **Plumbaginales**. Near to the **Caryophyllales**, especially to the *Portulacaceae* and *Basellaceae*, but the pollen morphology is different and seeds without perisperm and usually with mealy endosperm. Probably derived from the same stock as the **Caryophyllales**

Plumbaginaceae

Order 30: **Theligonales**. Affinities obscure. Some similiarities with the **Caryophyllales**, especially with the *Phytolaccaceae*, *Portulacaceae*, *Chenopodiaceae* and *Amaranthaceae*, but the pollen grain morphology is different and seeds without perisperm and with fleshy endosperm surrounding curved embryo. Possibly from the same stock as the **Caryophyllales**

Theligonaceae (*Cynocrambaceae*)

Subclass E: DILLENIIDAE

Superorder V: DILLENIANAE

Order 31: **Dilleniales**. A connecting link between the **Magnoliales** on the one hand, and the **Theales** and **Violales** on the other

Dilleniaceae

Crossosomataceae

Order 32: **Paeoniales**. Near to the **Dilleniales**, but distinguished by the structure of the thick-walled fleshy carpels, broad stigmas, the presence of a peculiar prominent lobed nectariferous disc (anatomically a part of the androecium), ovules with a massive outer integument which are borne on placental projections, the structure of the seed coat, and the extremely peculiar type of embryogeny

Paeoniaceae

Order 33: **Theales**. Near to the **Dilleniales**, and evidently derived from early *Dilleniaceae*. Nearest to the *Dilleniaceae* is the most primitive family of the order, the *Ochnaceae*

Ochnaceae

Lophiraceae (in many respects occupies, as it were, an intermediate position between the *Ochnaceae* and *Dipterocarpaceae*, and has much in common with the *Ancistrocladaceae*)

Dipterocarpaceae

Strasburgeriaceae (near to the *Ochnaceae*)

Ancistrocladaceae (related to the *Dipterocarpaceae*)

Dioncophyllaceae (the family relationships are not wholly clear, but many features, amongst them the structure of the actinocytic stomata, the structure of the petiole, and the morphology of the pollen grain wall apparently favour a position close to the *Ancistrocladaceae*)

Diegodendaceae (affinities obscure; probably related to the *Ochnaceae*)

Theaceae (including *Sladeniaceae*) (related to the *Ochnaceae* and shares with them a common origin from the **Dilleniales**; a basic group for the succeeding families)

Marcgraviaceae

Pentaphylacaceae

Tetrameristaceae
Caryocaraceae
Asteropeiaceae
Pellicieraceae
Quiinaceae
Medusagynaceae (systematic position is not wholly clear)
Oncothecaceae (affinities obscure)
Bonnetiaceae (in some respects an intermediate group between the *Theaceae* and *Clusiaceae*)
Clusiaceae (*Guttiferae sensu stricto*)
Hypericaceae (closely related to the *Clusiaceae*)
Elatinaceae (affinities uncertain; many similarities with the *Ochnaceae—Luxemburgieae* and *Hypericaceae*, also shows some similarities with the *Frankeniaceae*)

Order 34: **Violales (Cistales)**. Very near the **Theales**, and shares with them a common origin from the **Dilleniales**

Flacourtiaceae (including *Homaliaceae* and *Samydaceae*) (nearest to the *Dilleniaceae*)
Lacistemataceae (very near to the *Flacourtiaceae*)
Stachyuraceae (near to the *Flacourtiaceae*, especially to the *Scolopieae*)
Peridiscaceae (perhaps related to the *Flacourtiaceae*)
Violaceae (closely related to the *Flacourtiaceae* and is linked to them through the primitive tribe *Rinoreeae*)
Bixaceae (related to the *Flacourtiaceae*)
Cochlospermaceae (very near to the *Bixaceae*)
Cistaceae (related to the *Bixaceae* and *Cochlospermaceae* and shares with them a common origin from the *Flacourtiaceae*)
Scyphostegiaceae (probably derived from the *Flacourtiaceae*)

Order 35: **Passiflorales**. Derived from primitive **Violales** and so close to them that it is very difficult to draw a clear taxonomic boundary between the two most primitive families—*Passifloraceae* and *Flacourtiaceae*—of these orders

Passifloraceae (including the *Paropsieae*?, which in many respects are an intermediate group between the *Flacourtiaceae* and *Passifloraceae*)

Turneraceae (very near to the *Passifloraceae* and probably derived from them)

Malesherbiaceae (close to the *Turneraceae*)

Achariaceae (closely related to the *Passifloraceae*)

Caricaceae (related to the *Passifloraceae*, but much more advanced)

Order 36: **Cucurbitales**. Very closely related to the family *Passifloraceae* and scarcely merits separation into a distinct order **Cucurbitales**. However, according to C. Jeffrey, the monographer of this family, the *Cucurbitaceae* are rather isolated and stand remote from the *Passifloraceae*

Cucurbitaceae

Order 37: **Begoniales** (**Datiscales**). Probably derived from the **Violales**

Datiscaceae (including *Tetramelaceae*)

Begoniaceae

Order 38: **Capparales**. Derived from primitive representatives of the **Violales**, most likely from the *Flacourtiaceae*

Capparaceae (including *Cleomaceae* and *Oceanopapaver*) (related to the tribe *Oncobeae* of the family *Flacourtiaceae*. The Madagascan genus *Physaena* is very isolated within the *Capparaceae* and deserves a special study)

Koeberliniaceae (including *Canotiaceae*) (related to the *Capparaceae*)

Pentadiplandraceae (related to the *Capparaceae*)

Tovariaceae (related to the *Capparaceae*)

Moringaceae (related to the *Capparaceae*)
Brassicaceae (*Cruciferae*) (derived from the *Capparaceae–Cleomoideae* and linked to them through the primitive tribe *Stanleyeae*)
Resedaceae (probably related to and derived from the *Capparaceae*)
Emblingiaceae (affinities obscure)

Order 39: **Tamaricales**. Derived from the **Violales** (most probably from the *Flacourtiaceae*) and highly advanced
Tamaricaceae
Fouquieriaceae (probably related to the *Tamaricaceae*, but the sporoderm morphology is different and corolla is tubular)
Frankeniaceae

Order 40: **Salicales**. In all probability derived from the *Flacourtiaceae*, most likely from ancestors resembling present-day *Idesiinae*
Salicaceae

Superorder VI: ERICANAE

Order 41: **Ericales**. Derived from the **Theales**, with which it is closely linked through its most primitive families, *Saurauiaceae*, *Actinidiaceae* and *Clethraceae*
Saurauiaceae
Actinidiaceae (excluding *Sladenia*)
Clethraceae (close to the *Actinidiaceae*)
Ericaceae (close to the *Clethraceae*)
Epacridaceae (close to the *Ericaceae*, especially to the *Ericoideae*)
Pyrolaceae (close to the *Ericaceae*, especially to the *Rhododendroideae*)
Monotropaceae (close to the *Pyrolaceae*, but most likely by way of a common origin)

Cyrillaceae (evidently close relatives of the *Ericaceae*)
Empetraceae (close to the *Ericaceae*, especially the *Rhododendroideae*)
Grubbiaceae (probably related to the *Ericaceae*)

Order 42: **Diapensiales**. Evidently closest to the **Ericales**, especially to the *Epacridaceae*
Diapensiaceae

Order 43: **Ebenales**. Derived directly from the **Theales**. Especially close to the **Theales**, in particular to the family *Theaceae*, is the most primitive family of the order, the *Styracaceae*
Styracaceae (excluding *Afrostyrax*?)
Symplocaceae
Lissocarpaceae
Ebenaceae
Sapotaceae (including *Sarcospermataceae*?)

Order 44: **Primulales**. Stands fairly close to the **Ebenales** (especially to the *Styracaceae* and *Sapotaceae*) and all evidence points to a common origin with them from the *Theales*. The order *Primulales* is as a whole more advanced than the **Ebenales**
Myrsinaceae
Theophrastaceae
Primulaceae (including *Coridaceae*)

Superorder VII: MALVANAE

Order 45: **Malvales**. Evidently derived from early **Violales**. Exhibits many features in common with the *Flacourtiaceae* (especially the tribe *Scolopieae*). Evidently the gynoecium of the early **Malvales** was paracarpous
Elaeocarpaceae
Tiliaceae
Scytopetalaceae
Sarcolaenaceae

Rhopalocarpaceae (including *Sphaerosepalaceae*)

Sterculiaceae (very near to the *Tiliaceae*; to the *Sterculiaceae*, as well as to the *Tiliaceae*, are closely related the families *Bombacaceae* and *Malvaceae*, which are very near to each other)

Bombacaceae

Malvaceae

?*Huaceae* (including *Afrostyrax*? Affinities of the family are uncertain, but most probably with the *Sterculiaceae*)

Order 46: **Euphorbiales.** The primitive members of the *Euphorbiaceae* show obvious links with the **Malvales**, in particular with the family *Sterculiaceae*. Evidently the gynoecium was originally typically paracarpous in the **Euphorbiales**, as in the **Malvales**, and became multilocular as a result of fusion of intrusive placentae in the centre of the ovary (secondary syncarpy). On the other hand, the **Euphorbiales** have much in common with the primitive **Violales**, especially with the family *Flacourtiaceae*. One may therefore presume that the **Euphorbiales** arose from some ancient group intermediate between the *Flacourtiaceae* and **Malvales**

Buxaceae (including *Stylocerataceae*)

Simmondsiaceae (*Simmondsia*)

Daphniphyllaceae (probably belongs to this order and is related to the *Euphorbiaceae*)

Euphorbiaceae (a very heterogenous and evidently not wholly natural family)

Dichapetalaceae (near to the *Euphorbiaceae*, especially to the subfamily *Phyllanthoideae*)

Pandaceae (very near to the *Euphorbiaceae*)

Picrodendraceae (according to Novák, 1961, and G. Webster, personal communication, is related to the *Euphorbiaceae*)

Order 47: **Thymelaeales.** Has much in common with the *Euphorbiaceae* and also with the **Malvales** (especially with the *Tiliaceae* and the *Rhopalocarpaceae*) but closer to the former. All three orders have a common origin, but by all accounts the *Thymelaeales* are directly related to the ancestors of the **Malvales**. The **Thymelaeales** thus have a common origin with the **Euphorbiales** and **Malvales**, i.e. they arose from ancestors of the *Flacourtiaceae*-type

Thymelaeaceae (including *Aquilariaceae* and *Gonystylaceae*)

Subclass F: Rosidae

Superorder VIII: ROSANAE

Order 48: **Saxifragales** (incl. **Cunoniales** or **Grossulariales**).* Is linked to the **Dilleniales** through the *Cunoniaceae* and related families and evidently derived from the ancestors of the *Dilleniales*. A basic group for many other orders

Cunoniaceae

Davidsoniaceae (closely related to the *Cunoniaceae*)

Eucryphiaceae (related to the *Cunoniaceae*)

? *Paracryphiaceae* (affinities very obscure)

Crypteroniaceae

Brunelliaceae

Escalloniaceae (excluding *Polygonanthus* and including *Corokia*, which is closely related to *Argophyllum*. The genus *Carpodetus* is very isolated within the family)

Tribelaceae

Tetracarpaeaceae

Iteaceae

Brexiaceae

* Formerly I divided this order into two—the **Grossulariales** (**Cunoniales**) and **Saxifragales** *s.str.*

Phyllonomaceae (*Dulongiaceae*) (position uncertain)
Pterostemonaceae (position uncertain)
Grossulariaceae (palynologically an isolated family)
Hydrangeaceae (including *Philadelphaceae*)
Montiniaceae
Roridulaceae
Pittosporaceae
Byblidaceae
Bruniaceae
Penthoraceae
Crassulaceae
Cephalotaceae
Saxifragaceae
Vahliaceae
Francoaceae
Eremosynaceae
Parnassiaceae (including *Lepuropetalaceae*?)

Order 49: **Rosales**. Is linked to the **Dilleniales** and to the primitive families of the **Saxifragales** through the subfamily *Spiraeoideae* of the family *Rosaceae* and had a common origin with the *Saxifragales*. On the whole, more advanced than the **Saxifragales**; seeds usually without endosperm, and vessels usually with simple perforation

Rosaceae (including *Amygdalaceae* and *Malaceae*)
Chrysobalanaceae (excluding *Stylobasium*?)
Neuradaceae (probably related to the subfamily *Rosoideae* of the family *Rosaceae*. A very advanced family with a very peculiar structure of the pollen grain wall)

Order 50: **Fabales**. Close to the **Saxifragales** and evidently derived from their immediate ancestors. Consists of three closely related families which are treated by many botanists as subfamilies of a single family

Mimosaceae

Caesalpiniaceae
Fabaceae (*Papilionaceae*)

Order 51: **Connarales**. Has much in common with the **Saxifragales**, especially with the *Cunoniaceae*, but at the same time is also definitely linked with the **Fabales**. Evidently the **Connarales** have a common ancestry with the **Fabales** from the immediate ancestors of the **Saxifragales**

Connaraceae

Order 52: **Nepenthales**. Probably derived from the **Saxifragales**. The family *Droseraceae* is definitely related to the **Saxifragales** (especially to the *Parnassiaceae*), and the *Nepenthaceae* is very likely near to the *Droseraceae*

Droseraceae
Nepenthaceae

Order 53: **Podostemales**. Derived from the **Saxifragales**, most probably from the *Crassulaceae*-like ancestors

Podostemaceae

Superorder IX: MYRTANAE

Order 54: **Myrtales**. Evidently derived from the **Saxifragales**

Lythraceae
Sonneratiaceae
Punicaceae
Rhizophoraceae
Anisophylleaceae (including *Polygonanthaceae*)
Combretaceae
Lecythidaceae (including *Asteranthaceae*, *Barringtoniaceae*, *Foetidiaceae* and *Napoleonaceae*)
Myrtaceae (including *Heteropyxidaceae*)
Melastomataceae (including *Memecylaceae*)
Oliniaceae
Penaeaceae

Onagraceae
Trapaceae

Order 55: **Hippuridales** (**Haloragales**). Very close to the **Myrtales**, but are well distinguished by the presence of free styles (stylodia) and to a lesser degree by the absence of intraxylary phloem

Haloragaceae
Gunneraceae
Hippuridaceae

Superorder X: RUTANAE

Order 56: **Rutales**. In all probability derived from the **Saxifragales**

Anacardiaceae (including *Blepharocaryaceae*?)
Julianiaceae (very near to the *Anacardiaceae*)
Podoaceae (probably near to the *Anacardiaceae*)
Burseraceae
Simaroubaceae (including *Surianiaceae*?)
? *Stylobasiaceae* (*Stylobasium* is usually included in the *Chrysobalanaceae*, but, as has recently been shown by G. T. Prance, it is near to *Anacardiaceae* and related families and deserves a family status)
Rutaceae (including *Rhabdodendraceae*?)
Cneoraceae
Meliaceae (including *Ptaeroxylaceae*?)
? *Aitoniaceae* (*Nymania*) (affinities obscure)
? *Kirkiaceae* (affinities obscure)
? *Coriariaceae* (family relationships are not clear, and only provisionally included in the order **Rutales**)

Order 57: **Sapindales** (**Acerales**). Probably derived from the most primitive members of the **Rutales** or, rather, perhaps has a common origin with them from the **Saxifragales**

Staphyleaceae

Aceraceae
Sapindaceae
Akaniaceae
Hippocastanaceae
Bretschneideraceae
Melianthaceae
Greyiaceae (near to the *Melianthaceae*)
Sabiaceae (including *Meliosmaceae*; systematic position of the *Sabiaceae* is uncertain)

Order 58: **Geraniales**. Clearly connected with the **Rutales**, especially with the *Rutaceae*

Hugoniaceae
Linaceae
Ixonanthaceae (including *Irvingiaceae?*)
Humiriaceae
Erythroxylaceae
Lepidobotryaceae
Malpighiaceae (including *Ctenolophonaceae*)
Nitrariaceae
Zygophyllaceae
Balanitaceae
Peganaceae
Oxalidaceae (including *Averrhoaceae*)
Hypseocharitaceae (*Hypseocharis*)
Geraniaceae
Dirachmaceae
Vivianiaceae
Biebersteiniaceae
Tropaeolaceae
Balsaminaceae
Limnanthaceae (an isolated family, the affinities of which are not fully clear)

Order 59: **Polygalales**. Very closely linked with the **Geraniales**, especially with the family *Malpighiaceae*, which with almost equal reason might be included in either of these two orders

Trigoniaceae
Vochysiaceae
Polygalaceae (including *Diclidantheraceae* and *Xanthophyllaceae*)
Krameriaceae (relationships not fully clear, but most probably near to the *Polygalaceae*)
Tremandraceae (probably related to the *Polygalaceae*)

Superorder XI: ARALIANAE

Order 60: **Cornales** (**Umbellales**, **Apiales**, **Araliales**)*
Probably derived from the **Saxifragales**. The connection with the **Saxifragales** is shown by the morphological similarity of many primitive representatives of the **Cornales** to certain *Hydrangeaceae* and *Escalloniaceae*, and likewise by their anatomical resemblance
Cornaceae (excluding *Corokia*) (the genus *Griselinia* is very isolated within the family)
Garryaceae (close to the *Cornaceae* especially to the genera *Aucuba* and *Griselinia*)
Davidiaceae
Nyssaceae
Alangiaceae
Mastixiaceae
Helwingiaceae
Toricelliaceae
Araliaceae (the most primitive representatives of the family exhibit a relationship with the *Mastixiaceae*, *Helwingiaceae* and *Toricelliaceae* in the anatomy of the vegetative organs and in the structure of the gynoecium and ovules, as well as in the morphology of the sporoderm)
Apiaceae (*Umbelliferae*)

* Formerly I divided this order into two—the **Cornales** and **Araliales**. Recent studies on the *Mastixiaceae*, *Helwingiaceae* and *Toricelliaceae*, however, have shown that it is impossible to draw a distinct boundary between them.

Superorder XII: CELASTRANAE

Order 61: **Celastrales**. Evidently derived from the **Saxifragales**, most probably from the common ancestors of such families as the *Hydrangeaceae*, *Escalloniaceae* and *Brexiaceae*

Aquifoliaceae

Phellineaceae (related to the *Aquifoliaceae*, from which it differs, however, by having valvate petals, by the character of the inflorescence, the hemitropous or weakly campylotropous ovules, the wood anatomy, the quite different sporoderm morphology and the leaf venation)

Icacinaceae (including *Irvingbaileyaceae*)

Salvadoraceae

Celastraceae (including *Lophopyxidaceae*?) (*Campylostemonoideae*, *Cassinoideae* and *Tripterygioideae* are very isolated within the family)

Hippocrateaceae

Siphonodontaceae

Stackhousiaceae

Geissolomataceae (probably related to the *Celastraceae*)

Goupiaceae (relationships not fully clear)

Corynocarpaceae (relationships not fully clear)

? *Aextoxicaceae* (position uncertain)

Order 62: **Rhamnales**. Stands comparatively near to the **Celastrales**, but the stamens are opposite to the petals. The contemporary **Rhamnales** and **Celastrales** evidently had a common origin from diplostemonous ancestors

Rhamnaceae

Vitaceae

Leeaceae

Order 63: **Oleales**. Related to the **Celastrales** and in all probability had a common origin with them from the **Saxifragales**; especially closely connected

with such families as the *Aquifoliaceae*, *Icacinaceae* and *Salvadoraceae*
Oleaceae (including *Nyctanthes*)

Order 64: **Santalales**. The primitive representatives of the **Santalales** (especially the *Olacaceae*) are very near to the primitive families of the **Celastrales** (especially the *Icacinaceae*). However, the order **Santalales** does not appear to be a direct descendent of the order **Celastrales**, but rather shares a common origin with it
Olacaceae (including *Aptandraceae*)
Schoepfiaceae (*Schoepfia*)
Opiliaceae
Octoknemaceae
Erythropalaceae
Cardiopteridaceae (evidently related to the *Olacaceae*)
Santalaceae (near to and derived from the *Olacaceae*)
Dipentodontaceae (probably related to the *Olacaceae*)
Medusandraceae (relationships not fully clear, but evidently belong to the **Santalales**)
Misodendraceae (derived from the *Santalaceae*)
Loranthaceae (near to the *Santalaceae* and probably shares a common origin from the *Olacaceae*; including *Viscaceae*?)
Cynomoriaceae
Balanophoraceae (probably shares a common origin with the *Cynomoriaceae* from the *Santalaceae*)

Superorder XIII: PROTEANAE

Order 65: **Elaeagnales**. Probably had a common origin with the **Celastrales** and **Santalales** from the **Saxifragales**
Elaeagnaceae

Order 66: **Proteales**. Probably had a common origin with the **Elaeagnales**: exhibits definite links with the **Santalales** (a common origin)

Proteaceae

Subclass G: ASTERIDAE

Superorder XIV: LAMIANAE

Order 67: **Dipsacales**. Related to the **Cornales** (both morphological and serological data) and probably had a common origin with them. The *Caprifoliaceae* exhibits definite links both with the **Cornales** and with the *Hydrangeaceae* (especially with *Dichroa* and *Broussaisia*)

Caprifoliaceae (including *Alseuosmiaceae*, *Carlemanniaceae* and *Sambucaceae*?)

Adoxaceae

Valerianaceae

Dipsacaceae (including *Morinaceae* and *Triplostegiaceae*)

Order 68: **Gentianales**. A common origin with the **Dipsacales**

Desfontainiaceae (relationships not very clear)

Loganiaceae (including *Spigeliaceae* and *Strychnaceae*)

Antoniaceae (excluding *Peltanthera*)

Potaliaceae

Plocospermataceae (occupies in some respects an intermediate position between the *Loganiaceae* and *Apocynaceae*)

Apocynaceae

Asclepiadaceae (including *Periplocaceae*)

Gentianaceae

Menyanthaceae (relationships not altogether clear)

Dialypetalanthaceae (relationships not fully clear)

Rubiaceae (related to the *Loganiaceae* and allied families; including *Henriqueziaceae*? and *Naucleaceae*)

Order 69: **Polemoniales.** Very near to the **Gentianales** and probably derived from the immediate ancestors of the *Loganiaceae* and allied families

Polemoniaceae (including *Cobaeaceae*)

Convolvulaceae (including *Dichondraceae* and *Humbertiaceae*)

Cuscutaceae

Hydrophyllaceae (occupies in some respects an intermediate position between the *Polemoniaceae* and *Boraginaceae*, but stands much closer to the latter

Boraginaceae (including *Cordiaceae*, *Ehretiaceae* and *Wellstediaceae*)

Lennoaceae (related to the *Hydrophyllaceae* and *Boraginaceae*)

Hoplestigmataceae (related to the *Boraginaceae*)

Loasaceae (probably related to the *Boraginaceae* and *Hydrophyllaceae*)

Order 70: **Scrophulariales.** Very near to the **Polemoniales,** with which they had a common origin

Solanaceae (including *Goetzeaceae*)

Nolanaceae

Buddlejaceae (including *Peltanthera*) (related to the *Scrophulariaceae* and also show distinct relationships with the loganiaceous families)

Retziaceae (possibly related to the *Buddlejaceae* and *Scrophulariaceae*, but relationship is not very clear)

Scrophulariaceae

Bignoniaceae (very near to the *Scrophulariaceae* with which they probably had a common origin)

Pedaliaceae (near to the *Bignoniaceae* and probably derived from their immediate ancestors)

Trapellaceae (probably derived from the *Pedaliaceae*)

Martyniaceae

Gesneriaceae

Columelliaceae
Orobanchaceae
Lentibulariaceae
Myoporaceae (*Oftia* is very isolated within the family)
Globulariaceae
Acanthaceae (including *Mendonciaceae* and *Thunbergiaceae*)
Plantaginaceae (probably derived from the *Scrophulariaceae*)
Hydrostachyaceae (possibly had a common origin with the *Plantaginaceae*)

Order 71: **Lamiales**. Very near to and derived from the **Scrophulariales**

Verbenaceae (including *Avicenniaceae*, *Chloanthaceae* and *Symphoremaceae*)
Lamiaceae (including *Tetrachondraceae*) (very near to the *Verbenaceae*)
Phrymaceae (very near to the *Verbenaceae*)
Callitrichaceae (probably related to the *Lamiaceae* and *Verbenaceae*)

Superorder XV: ASTERANAE

Order 72: **Campanulales** (including **Goodeniales**). Has much in common with the **Polemoniales**. The orders **Campanulales** and **Polemoniales** have in all probability a common origin from the **Gentianales**

Campanulaceae (including *Pentaphragmataceae*)
Lobeliaceae (very near to the *Campanulaceae*)
Sphenocleaceae (near to the *Campanulaceae*)
Donatiaceae (related to the *Lobeliaceae*)
Stylidiaceae (near to the *Lobeliaceae* and *Donatiaceae*)
Goodeniaceae (probably derived from the *Lobeliaceae*)
Brunoniaceae (very near to the *Goodeniaceae*)

Order 73: **Calycerales**. Distinguished from the **Campanulales** mainly by the pendulous ovule. In having a pendulous ovule and certain embryological peculiarities the *Calyceraceae* approach the *Dipsacaceae*, but they differ from the latter in not having an epicalyx and in having an involucre of bracts surrounding the inflorescence, in the valvate or open aestivation of the corolla, in the more or less connate filaments, in the alternate or radical leaves, and in not having glandular hairs. The *Calyceraceae* cannot justifiably be referred to either the *Dipsacales* or to the *Campanulales*. The best solution of the problem is the separation of the *Calyceraceae* into a separate order, **Calycerales**. The order **Calycerales** stands nearest to the **Campanulales**, especially to the *Goodeniaceae*, which it definitely approaches in the structure of the pollen grain wall. Both these orders evidently had a common origin. At the same time, the **Calycerales** have much (and maybe even most) in common with the **Asterales**

Calyceraceae

Order 74: **Asterales**. Closest to the **Calycerales** but differs from them in the 2-lobed or 2-fid style, in the basal ovule, and in the absence of endosperm; it also has much in common with the **Campanulales**. All these three orders have a common origin from the **Gentianales** (or their immediate ancestors?)

Asteraceae (*Compositae*)

Class: LILIATAE or MONOCOTYLEDONES (monocots)

Subclass A: ALISMIDAE

Superorder I: ALISMANAE

Order 75: **Alismales**. Exhibits definite links with the

Nymphaeales among the dicots. The flowers are primitive, but the seeds are without endosperm and the pollen grains are 3-celled when shed

Butomaceae (*Butomus*)
Limnocharitaceae
Alismaceae (*Alismataceae*)

Order 76: **Hydrocharitales**. Near to and derived from the **Alismales**

Hydrocharitaceae

Order 77: **Najadales** or **Potamogetonales**. Near to the **Alismales** and probably derived from them

Scheuchzeriaceae
Juncaginaceae (including *Lilaeaceae*)
Aponogetonaceae
Zosteraceae
Posidoniaceae
Potamogetonaceae
Ruppiaceae
Zannichelliaceae
Cymodoceaceae (very near to the *Zannichelliaceae*)
Najadaceae

Subclass B: LILIIDAE

Superorder II: LILIANAE

Order 78: **Triuridales**. Despite the fact that the gynoecium is apocarpous and seeds have copious endosperm, the order as a whole is very specialised (small colourless saprophytes with leaves reduced to scales and with very small and generally unisexual flowers). Probably had a common origin with the **Alismales**. Has much in common with the lower representatives of the next order, especially with the *Liliaceae–Melanthioideae–Petrosavieae* (and *Tofieldieae*), but pollen grains are 3-celled when shed

Triuridaceae

Order 79: **Liliales**. The presence of endosperm and usually 2-celled pollen grains indicates that the order **Liliales** could not have originated from the order **Alismales**. Both orders, together with the **Triuridales**, have more probably a common origin from a hypothetical extinct group with endospermous seeds and 2-celled pollen grains, as in the **Liliales**, and an apocarpous gynoecium and numerous stamens, as in the **Alismales**. The subfamily *Melanthioideae* of the family *Liliaceae* is nearest to the ancestral type (and to the **Alismales** and **Triuridales**)

Liliaceae (including *Petrosaviaceae* and *Trilliaceae*)

Xanthorrhoeaceae (probably derived from the *Liliaceae–Asphodeloideae*)

Aphyllanthaceae (probably have a common origin with the *Xanthorrhoeaceae*)

Alliaceae (*Agapantheae*, *Allieae* and *Gilliesieae*)

Agavaceae (*Dracaenoideae* and *Agavoideae*)

Amaryllidaceae (probably have a common origin with the *Agavaceae* from the *Liliaceae*, most likely from the ancestors of contemporary *Hemerocalleae*)

Alstroemeriaceae (very near to the *Amaryllidaceae*, especially to the tribe *Haemantheae*)

Haemodoraceae (probably related to the *Liliaceae–Ophiopogonoideae*)

Hypoxidaceae (evidently near to the *Haemodoraceae*)

Velloziaceae (close to the *Hypoxidaceae*)

Philesiaceae (including *Petermanniaceae?*)

Tecophilaeaceae (related to the *Haemodoraceae*)

Cyanastraceae (probably have a common origin with the *Tecophilaeaceae*)

Asparagaceae (including *Ruscaceae*) (evidently derived from the *Liliaceae–Polygonateae*)

Smilacaceae (rather closely linked with the *Philesiaceae*)

Stemonaceae (including *Croomiaceae*) (exhibits some links with the *Philesiaceae* and *Tecophilaeaceae* and evidently also with the *Liliaceae*)

Dioscoreaceae (very near to the *Stemonaceae*, *Smilacaceae* and *Philesiaceae* and probably has a common origin with them from the *Liliaceae*)

Taccaceae (closely approaches the *Dioscoreaceae*)

Pontederiaceae (evidently related to the *Liliaceae*)

Philydraceae (evidently has a common origin with the *Pontederiaceae*)

Order 80: **Iridales**. Probably derived directly from the family *Liliaceae*, most likely from the subfamily *Melanthioideae*. The *Burmanniaceae* and *Corsiaceae* represent a parallel line with the **Orchidales**

Iridaceae

Geosiridaceae

Burmanniaceae (including *Thismiaceae*)

Corsiaceae

Order 81: **Zingiberales**. Probably derived directly from the **Liliales**; they have much in common with the *Amaryllidaceae* and especially with the subfamily *Asphodeloideae* of the family *Liliaceae*

Strelitziaceae

Musaceae

Heliconiaceae

Lowiaceae

Costaceae

Zingiberaceae

Cannaceae

Marantaceae

Order 82: **Orchidales**. Shows closest connections with the family *Hypoxidaceae* (**Liliales**), especially with the genera *Hypoxis* aad *Curculigo*. The connecting link between the *Hypoxidaceae* and *Orchidaceae* is the most primitive subfamily of the *Orchidaceae*, the *Apostasioideae*. In the most primitive genus of the *Apostasioideae*, *Neuwiedia*, there are three fertile stamens, the two lateral of the inner whorl (as in the subfamily *Cypripedioideae*) and one dorsal stamen of the outer whorl (as in all remaining orchids, except the genus *Satyrium*)

Orchidaceae

Subclass C: COMMELINIDAE

Superorder III: JUNCANAE*

Order 83: **Juncales**. Shows close connections with the *Xanthorrhoeaceae* and *Liliaceae*. Probably had a common origin with the **Bromeliales** and **Commelinales** from the *Liliaceae–Asphodeloideae*. Closest to the **Liliales** is the comparatively most primitive genus in the family *Juncaceae* and the order **Juncales**, the South African *Prionium*

Juncaceae

Thurniaceae (according to some recent authors perhaps not so nearly related to the *Juncaceae*, especially if one considers the results of anatomical studies)

Order 84: **Cyperales**. Evidently derived directly from the most primitive *Juncaceae*

Cyperaceae

Superorder IV: COMMELINANAE

Order 85: **Bromeliales**. Probably derived directly from the

* Formerly I referred the *Juncanae* to the *Liliidae*, but after conversations with Arthur Cronquist in September 1967 (at the conference on modern methods in plant taxonomy in Liverpool) I have come to the conclusion that they are more at home in the *Commelinidae*.

Liliales, most likely from the *Liliaceae–Asphodeloideae* (a common origin with the **Juncales** and **Commelinales**)
Bromeliaceae

Order 86: **Commelinales**. Probably has a common origin with the **Bromeliales**
Commelinaceae (including *Cartonemataceae*)
Mayacaceae
Xyridaceae (including *Abolbodaceae*?)
Rapateaceae

Order 87: **Eriocaulales**. Probably has a common origin with the **Commelinales**
Eriocaulaceae

Order 88: **Restionales**. Probably has a common origin with the **Commelinales**. From the evolutionary viewpoint, the most notable feature of the order **Restionales** is the appearance of characters which are characteristic of the grasses
Restionaceae (including *Anarthriaceae* and *Ecdeiocoleaceae*?)
Centrolepidaceae
Flagellariaceae
Hanguanaceae (systematic position not fully clear and it is not impossible that in the future it will be necessary to exclude it from the order **Restionales**)

Order 89: **Poales**. The grasses are the final link in a chain of anemophilous evolution, some basic stages of which may already be seen in the order **Commelinales**. By all accounts they arose directly from the **Restionales**, most likely from some extinct *Flagellariaceae* of the type of the living genus *Joinvillea*
Poaceae (*Gramineae*)

Subclass D: ARECIDAE

Superorder V: ARECANAE

Order 90: **Arecales**. Probably derived from immediate ancestors of the **Liliales**. Amongst the **Arecales** there are several genera with an apocarpous gynoecium (as, for example, *Trachycarpus*), and therefore they can have arisen only from the apocarpous ancestors of present-day *Liliaceae*

Arecaceae (*Palmae*)

Order 91: **Cyclanthales**. Evidently has a common origin with the **Arecales**.

Cyclanthaceae

Order 92: **Arales**. Most likely has a common origin with the **Arecales** and **Cyclanthales** from immediate ancestors of the **Liliales**

Araceae (*Acorus* is very isolated within the family)

Lemnaceae

Order 93: **Pandanales**. Stands closest to the **Cyclanthales**, with which it is linked through the comparatively most primitive genus *Freycinetia*

Pandanaceae

Order 94: **Typhales**. Possibly has a common origin with the **Pandanales**, but the affinity is not very clear

Sparganiaceae

Typhaceae (very near to the *Sparganiaceae*)

APPENDIX II

THE FLORISTIC REGIONS OF THE WORLD

The floristic regions here presented to the reader (Fig. 32, opp. p. 253) are based mainly on A. Engler and L. Diels (as revised by F. Mattick in Engler's *Syllabus der Pflanzenfamilien*, 1964), and on R. Good (1964), but with many modifications. The most important of these may be summarised as follows.

I have divided the Holarctic Kingdom into three subkingdoms (or dominions), namely the Boreal, the Tethyan or Ancient Mediterranean, and the Madrean (or Sonoran). The Boreal Subkingdom as I have defined it is a narrower concept than the Boreal Kingdom of Engler and of Good, since it does not include the Tethyan Subkingdom. My Tethyan Subkingdom differs from the Ancient Mediterranean Region of the late Russian systematist and phytogeographer, M. Popov, in that I have excluded from it south-west North America and Mexico and limited it to the Old World. On the other hand it is considerably larger than the Mediterranean Region in the strict sense, as it also includes the Macaronesian Region and the Irano-Turanian Region of Eig (which Good calls the 'Western and Central Asiatic Region').

I have considered it more correct to count the xerophilous floras of the south-west North American and Mexican uplands—Popov's 'Ancient Mediterranean of America'—as constituting a distinct subkingdom of the Holarctic, analogous in many respects to the Ancient Mediterranean Subkingdom and having much in common with it. I prefer to call this Subkingdom the Madrean, rather than the Sonoran (although this is just as appropriate), because the well-known palaeobotanists R. W. Chaney (1944) and D. I. Axelrod (1958) have already proposed the name Madro-Tertiary for the flora (geoflora) of its Tertiary fore-runner ('Madro' being taken from the name Sierra Madre), and this name has become widespread in the literature. On account of its floristic distinctiveness the contemporary Madrean flora deserves the rank of subkingdom.

In the Palaeotropical Kingdom I have added to the African, Indo-Malesian and Polynesian Subkingdoms two new ones—the Madagascan and the Neocaledonian. Both these subkingdoms are characterised by an extraordinarily high degree of endemism (about 85% in the Madagascan and 80% in the Neocaledonian), including a number of endemic families (especially distinctive and isolated in the case of Madagascar), and a large number of endemic genera. The flora of Madagascar (together with the Comoro, Seychelles and Mascarene Islands) is indeed so peculiar that perhaps it deserves to be ranked as a kingdom in its own right.

In comparison with the limits adopted in the *Syllabus*, I have found it necessary greatly to enlarge the area of the Papuan (Neoguinean) Region, which on my map extends from Celebes to the Solomon Islands. In the Polynesian Subkingdom I have distinguished only three regions—the Hawaiian, the Polynesian, and the Fijian. I have included the greater part of the Melanesian and Micronesian Regions in the Polynesian; but at the same time I have distinguished a new region—the Fijian. In revising the regional subdivision of the Indo-Malesian and Polynesian Subkingdoms I have been particulary assisted by the works of M. M. J. van Balgooy ('Preliminary plant-geographical analysis of the Pacific', in *Blumea* **10**, No. 2 (1960)) and R. F. Thorne ('Biotic distribution-patterns in the tropical Pacific', in *Pacific Basin Biogeography* (1963)), as well as by other sources, but I have not followed them in every detail.

In producing the map I have tried wherever possible to improve the accuracy of the regional boundaries. This was particularly necessary in the case of the Asian continent. For the boundary of the Irano-Turanian Region, the works of V. Grubov (*Plants of Central Asia*, pt. 1 (1963)), and many of the works of M. Zohary (in particular, *Plant Life of Palestine, Israel and Jordan* (1962)) have been especially useful. W. T. Stearn's '*Allium* and *Milula* in the Central and Eastern Himalayas' (*Bull. Br. Mus. nat. Hist., Bot.*, **2**, No. 6, 1960) and personal discussions with Dr M. Anantaswamy Rau in Dehra Dun have helped me in delimiting the western boundary of the Eastern Asian Region in the Himalayas. In defining the southern boundary of the Eastern Asian Region in China I have relied on the data given by An. A. Fedorov in the book *The Physical Geography of China* edited by V. T. Zaichikov (1964), and also in part on personal observations; while in drawing the southern boundary of this region in the southern islands of Japan I have been particularly helped by H. Hara's *An Outline of the Phytogeography of Japan* (1959), as well as by other sources. In the light of these works it has become necessary in particular to include within the Eastern Asian Region not only the Bonin (Ogasawara) Islands (as was clearly shown by van Balgooy), but also the Volcano (Kazan) Islands (Hara, 1959).

The Aleutian Islands must be included within the Euro-Siberian Region—see E. Hultén, *Flora of the Aleutian Islands*, etc. (1962). Good included them in the Pacific North American Region, while in the *Syllabus* they are included in the Arctic and Subarctic Region, which is quite incorrect.

As compared with the *Syllabus* and with Good's map, the regional boundaries in tropical Africa have also been considerably modified. In this case, I have used the works of R. W. J. Keay (*Vegetation Map of Africa* (1959)), Th. Monod (*Les Grandes Divisions Chorologiques de l'Afrique* (1957)) and also personal communication by Mr C. Jeffrey.

Last but not least, Dr A. Cronquist and Dr H. Irwin have kindly aided me in the drawing of boundaries in North, South and Central America.

I Holarctic Kingdom

Many *Magnoliaceae*, *Calycanthaceae* (!), many *Lauraceae*, the greater part of the *Ranunculaceae*, *Glaucidiaceae* (!), *Hydrastidaceae* (!), *Circaeasteraceae* (!), *Nandinaceae* (!), many *Berberidaceae*, the greater part of the *Papaveraceae*, *Hypecoaceae* (!), the greater part of the *Fumariaceae*, *Trochodendraceae* (!), *Tetracentraceae* (also in Upper Burma), *Cercidiphyllaceae* (!), *Eupteleaceae* (!), the greater part of the *Hamamelidaceae*, *Platanaceae* (!), *Eucommiaceae* (!), many *Ulmaceae*, *Cannabaceae*, many *Fagaceae*, the greater part of the *Betulaceae*,

Rhoipteleaceae (also in Tonkin), the greater part of the *Juglandaceae*, *Leitneriaceae* (!), the greater part of the *Caryophyllaceae*, many *Chenopodiaceae*, *Theligonaceae* (!), the greater part of the *Polygonaceae*, the greater part of the *Plumbaginaceae*, *Paeoniaceae* (!), *Crossosomataceae* (!), many *Theaceae*, many *Hypericaceae*, *Stachyuraceae* (!), many *Violaceae*, the greater part of the *Cistaceae*, *Koeberliniaceae* (!), the greater part of the *Brassicaceae*, *Resedaceae* and *Tamaricaceae*, *Fouquieriaceae* (!), the greater part of the *Frankeniaceae* and *Salicaceae*, many *Ericaceae*, the greater part of the *Pyrolaceae*, *Diapensiaceae* (also in Upper Burma), the greater part of the *Primulaceae*, *Simmondsiaceae* (!), many *Thymelaeaceae*, the greater part of the *Rosaceae*, *Pterostemonaceae* (!), the greater part of the *Hydrangeaceae*, *Penthoraceae* (!), many *Crassulaceae*, the greater part of the *Saxifragaceae* and *Parnassiaceae*, many *Fabaceae*, many *Onagraceae*, the greater part of the *Aceraceae* and *Hippocastanaceae*, *Bretschneideraceae* (!), many *Linaceae*, many *Zygophyllaceae*, *Peganaceae* (!), the greater part of the *Geraniaceae*, *Biebersteiniaceae* (!), *Limnanthaceae* (!), the greater part of the *Cornaceae*, *Davidiaceae* (!), *Toricelliaceae* (!), the greater part of the *Apiaceae*, many *Aquifoliaceae* and *Rhamnaceae*, *Dipentodontaceae* (also in Upper Burma), *Cynomoriaceae* (!), many *Oleaceae*, a great part of the *Gentianaceae*, many *Caprifoliaceae*, *Adoxaceae* (!) a great part of the *Valerianaceae*, *Dipsacaceae*, *Boraginaceae*, *Scrophulariaceae*, *Orobanchaceae*, *Plantaginaceae* and *Lamiaceae*, *Phrymaceae* (!), the greater part of the *Campanulaceae*, many *Asteraceae*, *Butomaceae s.str.* (!), *Scheuchzeriaceae* (!), the greater part of the *Liliaceae*, *Aphyllanthaceae* (!), the greater part of the *Alliaceae* and *Asparagaceae*, many *Iridaceae*, *Cyperaceae* and *Poaceae*.

A. Boreal Subkingdom

1. Arctic and Subarctic Region (Arctic Ocean coast and Islands, northern Norway, Faeroes, Iceland, Greenland, northern treeless parts of North America excluding the barren parts of the Alaskan peninsula, the North American Arctic Islands, Pribilof Islands). No well defined endemic genera and only a few endemic species of *Ranunculus*, *Braya* (*Brassicaceae*), *Salix*, *Pedicularis*, *Chrysanthemum*, *Nardosmia* and various genera of grasses.

2. Euro-Siberian Region (Western and Central Europe,

Scandinavia, Eastern Europe, Danube basin, Caucasus, northern Anatolia, Hyrcania, western Siberia, Altai-Transbaikalia, north-eastern Siberia, Kamchatka, Aleutian Islands). Many endemic genera of various families and numerous endemic species.

3. Eastern Asian (Sino-Japanese) Region (the eastern Himalayas extending from western Nepal, Upper Assam, Continental China except tropical and arid parts, a part of Taiwan, Korea, south-eastern Siberia or Primorie, southern and central Sakhalin, Kunashir, Shikotan and Iturup of the Kuril Islands, Hokkaido, Honshu, Shikoku, Kyushu, Ryukyu Islands (except Sakishima Group), Bonin Islands and Volcano or Kazan Islands). Endemic families: *Glaucidiaceae*, *Circaeasteraceae*, *Nandinaceae*, *Trochodendraceae*, *Tetracentraceae* (also in Upper Burma), *Cercidiphyllaceae*, *Eupteleaceae*, *Eucommiaceae*, *Rhoipteleaceae* (also in Tonkin), *Stachyuraceae*, *Bretschneideraceae*, *Davidiaceae*, *Toricelliaceae*, *Dipentodontaceae* (also in Upper Burma) and *Trapellaceae*. More than 300 endemic genera and numerous endemic species.

4. Atlantic North American Region. Endemic families *Hydrastidaceae* and *Leitneriaceae*, no fewer than a hundred endemic genera (including *Sanguinaria*, *Sarracenia*, *Dirca*, *Neviusia*, *Hudsonia*, *Dionaea*, *Yeatesia*, *Pleea*, *Uvularia*) and numerous endemic species.

4a. Canadian-Appalachian Subregion (from Alaska eastwards and southwards to Ontario and Quebec and from Ontario and Quebec southwards across the eastern states to northern Georgia, northern Louisiana, and eastern Texas). Strong general resemblance to the Euro-Siberian flora.

4b. Subregion of Southern Atlantic North America (Coastal Plain of the Atlantic and Gulf States and the Mississipi embayment). Well marked similarities to the Sino-Japanese flora.

4c. Central Grasslands Subregion. The Prairies and Plains east of the Rocky Mountains.

5. Rocky Mountain Region (mountains of the western states of the U.S.A. and western Canada from Alaska to California and New Mexico). Few endemic genera (*Tolmiea*, *Sidalcea*, etc.), but many endemic species.

B. Tethyan (Ancient Mediterranean) Subkingdom

6. Macaronesian Region (Azores, Madeira, Canaries, Cape Verdes). Comparatively few endemic genera (about thirty), about half of which are confined to the Canaries, but numerous endemic species amongst them *Laurus azorica*, *Apollonias barbusana*, *Persea indica*, *Ocotea foetens*, *Clethra arborea*, *Arbutus azorica*, *Pittosporum coriaceum*, *Sideroxylon marmulano*, *Ilex canariensis*, *Sambucus maderensis*, *Viburnum rugosum*, arborescent or frutescent spp. of *Sempervivoideae*, *Statice*, *Euphorbia*, *Cneorum*, *Echium*, *Sonchus* etc., *Dracaena draco*, *Phoenix jubae*.

7. Mediterranean Region (the larger part of the Iberian peninsula, north Mediterranean coasts and islands, Morocco, north Algeria, Tunisia, north-western Tripolitania, Cyrenaica, north-west Egypt, a large part of Palestine, Lebanon, western Syria, west and south Anatolia). One endemic monotypic West Mediterranean family *Aphyllanthaceae* (*Liliales*), many endemic genera and species. The most characteristic are: *Laurus nobilis*, the evergreen oaks *Quercus ilex* and *Q. coccifera*, *Cistus* spp., *Helianthemum* spp., *Arbutus unedo* and *A. andrachne*, *Erica* spp., *Myrtus communis*, *Nerium oleander*, *Vitex agnus-castus*, numerous *Apiaceae*, *Boraginaceae*, *Lamiaceae*, *Asteraceae*, *Liliaceae*, *Amaryllidaceae* and *Poaceae*. Small and modified exclaves of the Mediterranean flora are found in the southern mountainous part of the Crimea and along the Black Sea coast in north Anatolia and in the Caucasus.

8. Irano-Turanian (Western and Central Asiatic) Region (a part of Palestine and Syria, north Iraq, Inner Anatolia, parts of south Transcaucasia, the Iranian Highlands except the Caspian coast of Iran and certain parts of south Iran and Baluchistan in Western Pakistan, deserts of south Russia, Transcaspia, Turkestan, Karakoram, Mongolia, Sin Kiang, Nan Shan, Ala Shan, Ordos Plateau, Tsaidam, Tibetan Plateau). No endemic family, but numerous endemic genera and species (mainly *Chenopodiaceae*, *Caryophyllaceae*, *Brassicaceae*, *Rosaceae*, *Fabaceae*, *Zygophyllaceae*, *Scrophulariaceae*, *Lamiaceae*, *Campanulaceae*, *Asteraceae*, *Liliaceae* and *Poaceae*). Many endemic species of the genera *Salsola*, *Calligonum*, *Atraphaxis*, *Astragalus*, *Oxytropis*, *Ferula*, *Verbascum*, *Salvia*, *Phlomis*, *Eremostachys*,

Cousinia, *Centaurea*, *Artemisia*, *Eremurus*, *Allium*. There are three subregions within the Irano-Turanian Region: Armeno-Iranian (Irano-Anatolian), Mesopotamian, and Centralasiatic. An outlier (exclave) of the Irano-Turanian flora occupies a considerable area in North Africa, notably in its western part ('Mauritanian Steppes').

C. Madrean (Sonoran) Subkingdom

9. Madrean (Sonoran) Region (most of Nevada and Utah and parts of adjacent states, California, extending into south-western Oregon and northern Lower California, hot deserts and valleys from southern California to Arizona, New Mexico and Texas and southward into Mexican Highlands). Endemic families *Crossosomataceae*, *Koeberliniaceae*, *Fouquieraceae*, *Simmondsiaceae* and *Pterostemonaceae* and many endemic genera (including *Umbellularia*, *Romneya*, *Dendromecon*, *Dasylirion*, *Beschorneria*, *Sarcodes*).

II Palaeotropical Kingdom

Many *Annonaceae*, *Amborellaceae* (!), many *Lauraceae* and *Piperaceae*, *Barclayaceae* (!), *Nepenthaceae* (also in N. Australia), *Didymelaceae* (!), many *Hamamelidaceae*, *Myrothamnaceae* (!), many *Moraceae* and *Urticaceae*, *Barbeyaceae* (!), the greater part of the *Fagaceae*, *Balanopaceae* (also in tropical Australia), the greater part of the *Molluginaceae*, many *Aizoaceae*, *Didiereaceae* (!) many *Chenopodiaceae*, *Lophiraceae* (!), *Dipterocarpaceae* (!), *Strasburgeriaceae* (!), *Ancistrocladaceae* (!), *Dioncophyllaceae* (!), many *Theaceae*, *Pentaphylacaceae* (!), *Tetrameristaceae* (!), *Asteropeiaceae* (!), *Medusagynaceae* (!), many *Clusiaceae* and *Flacourtiaceae*, *Scyphostegiaceae* (!), many *Passifloraceae*, *Cucurbitaceae* and *Capparaceae*, *Pentadiplandraceae* (!), *Moringaceae* (!), the greater part of the *Ebenaceae*, many *Sapotaceae*, many *Myrsinaceae*, *Elaeocarpaceae* and *Tiliaceae*, *Scytopetalaceae* (!), *Sarcolaenaceae* (!), *Rhopalocarpaceae*, many *Sterculiaceae*, *Bombacaceae*, *Malvaceae*, *Euphorbiaceae* and *Thymelaeaceae*, *Crypteroniaceae* (!), *Montiniaceae* (!), *Vahliaceae* (!), many *Mimosaceae*, *Caesalpiniaceae* and *Fabaceae*, the greater part of the *Connaraceae*, many *Combretaceae*, *Lecythidaceae*, *Myrtaceae* and *Melastomataceae*, *Oliniaceae* (!), many *Anacardiaceae*, *Burseraceae*, *Rutaceae* and *Meliaceae*, *Kirkiaceae* (!), many *Sapindaceae*,

Melianthaceae (!), *Lepidobotryaceae* (!), *Balanitaceae* (!), many *Oxalidaceae*, *Dirachmaceae* (!), the greater part of the *Mastixiaceae*, many *Araliaceae* and *Aquifoliaceae*, *Phellineaceae* (!), *Salvadoraceae* (!), many *Celastraceae*, *Rhamnaceae* and *Vitaceae*, *Leeaceae* (!), *Octoknemaceae* (!), *Erythropalaceae* (!), *Medusandraceae* (!), many *Loranthaceae*, *Proteaceae*, *Rubiaceae*, and *Convolvulaceae*, *Hoplestigmataceae*, many *Gesneriaceae*, the greater part of the *Acanthaceae*, many *Asteraceae*, *Cyanastraceae* (!), many *Dioscoreaceae*, *Geosiridaceae* (!), *Musaceae* (!), *Lowiaceae* (!), many *Orchidaceae* and *Eriocaulaceae*, *Hanguanaceae* (!), *Flagellariaceae* (!), many *Poaceae*, *Arecaceae*, *Araceae*, and *Pandanaceae*.

A. African Subkingdom

10. Saharo-Sindian Region (from the Atlantic coast of North Africa through the Sahara, the Sinai Peninsula, most of Arabia except the south, more than half of Palestine, part of Syria, south Iraq, south Iran, part of southern Baluchistan, Sind in Western Pakistan and Rajasthan in India). No endemic families and few really endemic genera, among them *Nucularia*, *Fredolia*, *Agathophora*, *Zilla*, *Ochradenus*, *Neurada*, *Rhanterium*. The flora is comparatively poor.

11. Sudano-Angolan Region (Senegal eastwards to Sudan, north-east and east tropical Africa, Socotra, south-west Arabian peninsula, Mozambique, Zambia, Malawi, Rhodesia, Transvaal, Natal, most of Angola, parts of Bechuanaland, South-West Africa, Orange Free State and Cape Province). Few endemic families (e.g., *Kirkiaceae*, *Barbeyaceae*) and genera, but a large number of endemic species. The montane areas have a discontinuous distribution and are occupied by a distinct Afromontane flora. Apart from this, four subregions may be recognised.

11a. Sahelo-Sudanian Subregion (Senegambia eastwards to Sudan, parts of Uganda and western Kenya). Species of *Acacia* and *Commiphora* and (in the less arid parts) *Isoberlinia* are especially characteristic of this subregion, which is floristically poor, especially in endemics.

11b. Somalo-Ethiopian Subregion, extending southwards into central Kenya and Tanzania; lowland areas characterised by species of *Acacia* and *Commiphora* but with a considerable number of associated endemic genera (e.g. *Barbeya*, *Sevada*, *Carania*, *Cephalopentandra*, *Socotora*) and species. In its northern part this subregion includes a large Afromontane area characterised by species of *Olea*, *Ocotea*, *Juniperus*, *Podocarpus*, *Hagenia abyssinica*, and species of *Alchemilla* and *Helichrysum*.

11c. South Tropical African Subregion (Tanzania, southern Congo, Angola, Zambia, Malawi, Mozambique, Rhodesia, Transvaal, Natal); characteristic are species of *Brachystegia*, *Julbernardia*, *Acacia*, *Commiphora*, *Terminalia*, *Combretum* and *Pleiotaxis*, also *Kirkia acuminata*, *Baikiaea plurijuga* and *Colophospermum mopane*; in the south-east a considerable area of grassland with abundant *Themeda triandra* occurs which is related to the grasslands of the Afromontane flora.

11d. Kalaharian Subregion (Bechuanaland and part of South-West Africa), characterised by abundant *Acacia* and *Commiphora* with many endemic species.

In conclusion it is also necessary to mention the presence in the coastal regions of East Africa of a very narrow and discontinuous area characterised by a considerable number of endemics and showing some generic affinities with the Madagascan Region (e.g., *Ludia*, *Aphloia*, *Brexia*, *Macphersonia*, *Mascarenhasia*).

12. West African Rain-forest Region (Upper Guinea, Cameroun and Islands, Gabon, Congo Basin). A few endemic families (*Dioncophyllaceae*, *Scytopetalaceae*, *Octoknemaceae*, *Medusandraceae*, *Hoplestigmataceae*), many endemic genera and numerous endemic species.

13. Namib-Karroo Region (western part of South-West Africa and the Karroo). Very many endemic species; *Mesembryanthemum* and allied genera, *Tetragonia*, *Pelargonium*, *Rhigozum*, *Pentzia*, *Pteronia* and other shrubby *Asteraceae* are especially characteristic; in the northern parts of the Namib *Welwitschia* and *Acanthosicyos horridus* are endemic.

14. Region of Ascension and St Helena. Only five endemic genera (*Mellissia*, *Nesiota*, *Commidendrum*, *Melanodendron* and *Petrobium*), all from St Helena. Ascension Island has only some eight indigenous species, of which at least two, *Euphorbia origanoides* and *Hedyotis adscensionis*, are endemic.

B. Madagascan Subkingdom

15. Madagascan Region (Madagascar and the Comoros, Aldabra, the Seychelles, the Mascarenes). Eight endemic families (*Didymelaceae*, *Didiereaceae*, *Diegodendraceae*, *Asteropeiaceae*, *Medusagynaceae* only in Seychelles, *Sarcolaenaceae*, *Rhopalocarpaceae*, *Geosiridaceae*) and about 300 endemic genera, the great majority of which are found on Madagascar itself.

C. Indo-Malesian (Indo-Malayan) Subkingdom

16. Indian Region (Malabar coast and southern India, Deccan, Ganges Plain, flanks of the Himalayas, Ceylon). No endemic families and no more than 150 endemic genera, of which the great majority are monotypic and very local.

17. Indo-Chinese (Continental South-east Asian) Region (eastern Assam, Burma, Indo-China and Thailand, the Andamans and Nicobars, south China, Hainan, a greater part of Taiwan, Sakishima Islands). No endemic families, but more than 250 endemic genera.

18. Malesian (Malayan) Region (the Malay Peninsula, Java, Sumatra and the Sunda Islands, Borneo, Philippines, Botel Tobago). A few endemic families (*Tetrameristaceae* and *Scyphostegiaceae*) and many endemic genera.

19. Papuan Region (Celebes, Tahuna, Talaud, Moluccas, Banda, Tanimbar (?), Aru, Misoöl, Salawati, Waigeo, New Guinea, D'Entrecasteaux, Trobriand Murna, Louisiade, Bismark, Admiralty, Solomon Islands). No endemic families and not many endemic genera, but the proportion of endemic species is high.

D. Polynesian Subkingdom

20. Hawaiian Region (Hawaii, Johnston). No endemic families, but about 20% of genera are endemic (among them some woody members of the *Lobeliaceae*). More than 90% of the native Hawaiian species are endemic.

21. Polynesian Region (Mariana, Palau, Caroline, Marcus, Wake, Marshall, Nauru, Ocean, Gilbert, Howland, Baker, Ellice, Phoenix, Wallis, Horn, Tokelau, Line, Nike, Cook, Society, Austral, Rapa, Tuamotu, Marquesas, Gambier, Pitcairn, Henderson, Easter,

Sala-y-Gomez). No endemic families and comparatively few endemic genera.

22. Fijian Region (Santa Cruz, Banks, New Hebrides, Fiji, Tonga, Samoa). A remarkable monotypic endemic family *Degeneriaceae* and about fifteen endemic genera.

E. Neocaledonian Subkingdom

23. Neocaledonian Region (New Caledonia, Isle of Pines, Loyalty). Five endemic families (*Amborellaccae*, *Strasburgeriaceae*, *Oncothecaceae*, *Paracryphiaceae* and *Phellineaceae*) and about 100 endemic genera (among them *Exospermum*, *Canacomyrica*, *Oceanopapaver*, *Maxwellia*, *Memecylanthus*, *Periomphale*, *Serresia* and *Solmsia*). Over 80% of the species are endemic.

III Neotropical Kingdom

Many *Annonaceae*, *Gomortegaceae* (!), *Lactoridaceae* (!), many *Lauraceae*, *Piperaceae*, *Moraceae*, *Urticaceae*, the greater part of the *Phytolaccaceae*, the greater part of the *Nyctaginaceae*, the greater part of the *Cactaceae*, *Halophytaceae* (!), many *Chenopodiaceae* and *Ochnaceae*, *Theaceae*, *Marcgraviaceae* (!), *Caryocaraceae* (!), *Pellicieraceae* (!), *Quiinaceae* (!), many *Clusiaceae* and *Flacourtiaceae*, *Lacistemataceae* (!), *Peridiscaceae* (!), *Bixaceae* (!), *Cochlospermaceae* (also in southern Arizona), many *Passifloraceae*, *Malesherbiaceae* (!), *Caricaceae* (except the tropical African genus *Cylicomorpha*), many *Cucurbitaceae*, the greater part of the *Begoniaceae*, many *Capparaceae*, *Tovariaceae* (!), *Cyrillaceae* (also in North America), *Lissocarpaceae* (!), many *Sapotaceae* and *Myrsinaceae*, *Theophrastaceae* (also in Hawaii), many *Elaeocarpaceae*, *Tiliaceae* and *Sterculiaceae*, the greater part of the *Bombacaceae*, many *Malvaceae* and *Euphorbiaceae*, *Brunelliaceae* (!), *Francoaceae* (!), many *Mimosaceae*, *Caesalpiniaceae* and *Fabaceae*, many *Connaraceae*, *Lythraceae*, *Combretaceae*, *Lecythidaceae*, *Myrtaceae*, *Melastomataceae* and *Anacardiaceae*, *Julianiaceae* (!), many *Burseraceae* and *Meliaceae*, *Picrodendraceae* (!), many *Sapindaceae*, *Humiriaceae* (with one species of *Saccoglottis* in tropical West Africa), the greater part of the *Erythroxylaceae*, the greater part of the *Malpighiaceae*, many *Oxalidaceae*, *Vivianiaceae* (!),

Vochysiaceae (1 monotypic genus *Erismadelphus* in tropical West Africa), many *Araliaceae*, *Aquifoliaceae* and *Celastraceae*, *Goupiaceae* (!), *Aextoxicaceae* (!), many *Rhamnaceae* and *Vitaceae*, many *Loranthaceae* and *Proteaceae*, *Desfontainiaceae* (!), *Plocospermataceae* (!), *Dialypetalanthaceae* (!), many *Rubiaceae* and *Convolvulaceae*, *Loasaceae* (also 1 species of *Kissenia* in Africa and Arabia), the greater part of the *Solanaceae*, *Nolanaceae* (!), many *Bignoniaceae* and *Gesneriaceae*, *Columelliaceae* (!), many *Acanthaceae*, *Calyceraceae* (!), many *Asteraceae*, *Alstromeriaceae* (!), *Bromeliaceae* (1 species in tropical West Africa), *Heliconiaceae* (!), *Cannaceae* (!), many *Orchidaceae*, *Thurniaceae* (!), many *Eriocaulaceae*, *Poaceae* and *Arecaceae*, *Cyclanthaceae* (!), many *Araceae*.

24. Caribbean Region (southern Lower California, Mexican lowlands and coast, Revilla Gigedo, Clipperton, southern Florida and the Florida Keys, West Indies, Bahamas, Bermudas, Guatemala-Panama, north Colombia, north Venezuela and Trinidad). Two endemic families (*Picrodendraceae* and *Plocospermataceae*) and more than 500 endemic genera. There are no endemic genera on the Bermudas. A very high proportion of endemic species.

25. Region of Venezuela and Surinam (Orinoco Basin, Uplands of Venezuela and the sandstone mountains of British Guiana and the Tafelberg in Surinam). No endemic families and most probably fewer than 100 endemic genera.

26. Amazon Region (lowlands of the Amazon basin and eastern coasts of Brazil). Only one endemic family (*Dialypetalanthaceae*), but many endemic tribes and about five hundred endemic genera (among them *Henriquezia*, *Platycarpum* and *Duckeodendron*). At least 3000 endemic species.

27. Central Brazilian Region (Uplands of central Brazil, Highlands of eastern Brazil, Grand Chaco, St. Paul Rocks, Fernando de Noronha, Martin Vaz). No endemic families, but about four hundred endemic genera (*Antonia* and *Diclidanthera* among them).

28. Pampas Region (Uruguay and south-eastern Brazil, Argentine pampas, western Argentina). One endemic family (*Halophytaceae*). Probably no more than fifty endemic genera.

29. Andean Region (Galapagos, Cocos, flanks of the Andes, Montane Andes, Atacama Desert, Chilean sclerophyll Zone). Six

endemic families—*Gomortegaceae* (Chile), *Malesherbiaceae* (also in Argentina), *Francoaceae* (Chile), *Aextoxicaceae* (Chile), *Desfontainiaceae* (Chile and Peru), *Nolanaceae* (northern Chile and Peru). Probably several hundred endemic genera (*Lardizabala*, *Berberidopsis*, *Lapageria* and *Jubaea* among them).

30. Fernandezian Region (Juan Fernandez, Desventurados). One almost extinct monotypic endemic family *Lactoridaceae* and about fifteen endemic genera (among them *Centaurodendron*, *Dendroseris* and a palm genus *Juania*). About 70% of the 143 species are endemic, among them three arborescent species of *Chenopodium*.

IV Cape Kingdom

Many *Aizoaceae* and *Ericaceae*, *Grubbiaceae* (!), many *Euphorbiaceae* and *Fabaceae*, *Roridulaceae* (!), *Bruniaceae* (!), many *Crassulaceae*, *Penaeaceae* (!), *Greyiaceae* (!), *Geissolomataceae* (!), many *Proteaceae*, *Retziaceae* (!), many *Asclepiadaceae*, *Rubiaceae*, *Lamiaceae*, *Solanaceae*, *Scrophulariaceae*, *Asteraceae*, *Liliaceae*, *Amaryllidaceae*, *Iridaceae*, *Orchidaceae*, *Restionaceae*.

31. Cape Region (the coast zone from Clanwilliam on the west to the neighbourhood of Port Elisabeth on the east). Very rich flora with seven endemic families and numerous endemic genera and species.

V Australian Kingdom

Austrobaileyaceae (!), the greater part of the *Casuarinaceae*, *Gyrostemonaceae* (!), many *Aizoaceae*, many *Chenopodiaceae*, *Emblingiaceae* (!), the greater part of the *Epacridaceae*, many *Cunoniaceae*, *Tetracarpaeaceae*, (!), the greater part of the *Pittosporaceae*, *Byblidaceae* (!), *Cephalotaceae* (!), *Eremosynaceae* (!), many *Myrtaceae*, many *Rutaceae*, the greater part of the *Stackhousiaceae*, *Akaniaceae* (!), *Tremandraceae* (!), many *Proteaceae*, the greater part of the *Myoporaceae*, *Stylidiaceae* and *Goodeniaceae*, *Brunoniaceae* (!), many *Fabaceae* and *Mimosaceae*,

Euphorbiaceae, *Asteraceae*, *Xanthorrhoeaceae*, *Orchidaceae*, *Cyperaceae*, *Poaceae*. About 400 endemic genera.

32. North and East Australian Region (northern forests, Queensland forests, south-eastern forests, Tasmania). Three endemic families (*Austrobaileyaceae*, *Tetracarpaeaceae* and *Akaniaceae*) and more than 150 endemic genera. Tasmania has 10 endemic genera (*Isophysis* and *Prionotes* among them).

33. South-west Australian Region. Three endemic families (*Cephalotaceae*, *Eremosynaceae* and *Emblingiaceae*) and about 125 endemic genera (including *Dryandra*, *Nuytsia*, *Stirlingia*, etc.). A very high degree of endemism on the species level (75% or more).

34. Central Australian Region (north and east savannas, central deserts, South Australia). No endemic families and about forty endemic genera (many of them in the families *Chenopodiaceae*, *Brassicaceae* and *Asteraceae*).

VI Antarctic Kingdom

Hectorellaceae (!), *Tribelaceae* (!), *Misodendraceae* (also in Pampas Region), *Donatiaceae* (!), and about forty endemic genera (*Laurelia*, *Oncostylus*, *Tetrachondra* and *Rostkovia* among them).

35. New Zealand Region (North Island, South Island, Lord Howe, Norfolk, Kermedec, Chatham, Bounty, Antidopes, Campbell, Auckland). No endemic families and only about thirty endemic genera (among them *Ixerba*, *Carlmichaelia*, *Alseuosmia*, *Negria*, and *Phormium* and *Rhopalostylis*).

36. Patagonian Region (Patagonia and Fuegia, southern Andes, Falkland). One endemic family *Tribelaceae* and a few endemic or almost endemic genera, among them *Saxifragella*, *Niederleinia*, *Magallana*, *Lebetanthus*, *Monttea*).

37. Region of the South Temperate Oceanic Islands (South Georgia, Tristan da Cunha with Nightingale Island, Prince Edward and Marion, Crozet, Kerguelen, Heard and Macdonald, Amsterdam and St. Paul, Macquarie). No endemic families and only two endemic genera—*Lyallia* on Kerguelen and *Pringlea* on Kerguelen and Crozet islands.

BIBLIOGRAPHY

ABEL, O. 1929. *Paläobiologie und Stammesgeschichte*. Jena.

ALEXANDROV, A. G. 1954. *Plant anatomy*. Moscow. (In Russian.)

ANDREWS, N. 1961. *Studies in paleobotany*. New York and London.

ARBER, A. 1925. *Monocotyledons: a morphological study*. Cambridge.

ARBER, A. 1937. The interpretation of the flower: a study of some aspects of morphological thought. *Biol. Rev.*, **12**, 157-184.

ARBER, E. A. N. and PARKIN, J. 1907. On the origin of angiosperms. *J. Linn. Soc. Bot.*, **38**, 29-80.

ARLDT, TH. 1907. *Die Entwicklung der Kontinente und ihrer Lebewelt*. Leipzig.

ARLDT, TH. 1938. *Die Entwicklung der Kontinente und ihrer Lebewelt* (2nd ed.). I. Berlin.

ARNOLD, C. A. 1947. *An introduction to paleobotany*. New York.

ASAMA, K. 1960. Evolution of the leaf forms through the ages explained by the successive retardation and neoteny. *Sci. Rep. Tôhoku Univ.*, Ser. 2, Spec. vol. (4), 252-280.

ASANA, J. J. and ADATIA, R. D. 1947. Contributions to the embryology of the *Anonaceae*. *J. Univ. Bombay*, Sect. A (Biol. Sci.), **16** (3), 7-21.

AXELROD, D. I. 1950. Classification of the Madro-Tertiary flora. *Publs. Carnegie Instn.*, **590**, 1-22.

AXELROD, D. I. 1952. A theory of angiosperm evolution. *Evolution*, **4**, 29-60.

AXELROD, D. I. 1958. Evolution of the Madro-Tertiary geoflora. *Bot. Rev.*, **24**, 433-509.

AXELROD, D. I. 1959. Poleward migration of early angiosperm flora. *Science*, **130** (3369), 203-207.

AXELROD, D. I. 1960. The evolution of flowering plants. In *Evolution after Darwin*, **1**, 227-305.

AXELROD, D. I. 1961. How old are the angiosperms? *Am. J. Sci.*, **259**, 447-459.

AXELROD, D. I. 1966. Origin of deciduous and evergreen habits in temperate forest. *Evolution*, **20**, 1-15.

BAIKOVSKAYA, T. N. 1956. The Upper Cretaceous floras of Northern Asia. *Acta. Inst. Bot. Acad. Sci. USSR*, ser. VIII, Palaeobotany, **2**, 47-194. (In Russian.)

Bailey, I. W. 1944a. The development of vessels in angiosperms and its significance in morphological research. *Am. J. Bot.*, **31**, 421-428.

Bailey, I. W. 1944b. The comparative morphology of the *Winteraceae*. III. Wood. *J. Arnold Arbor.*, **25**, 97-103.

Bailey, I. W. 1949. Origin of the angiosperms: need for a broadened outlook. *J. Arnold Arbor.*, **30**, 64-70.

Bailey, I. W. 1956. Nodal anatomy in retrospect. *J. Arnold Arbor.*, **37**, 269-287.

Bailey, I. W. 1957. Additional notes on the vesselless dicotyledon, *Amborella trichopoda* Baill. *J. Arnold Arbor.*, **38**, 374-378.

Bailey, I. W. and Howard, R. A. 1941. The comparative morphology of the *Icacinaceae*. II. Vessels. IV. Rays of the secondary xylem. *J. Arnold Arbor.*, **22**, 171-187; 556-568.

Bailey, I. W. and Nast, C. G. 1943a. The comparative morphology of the *Winteraceae*. I. Pollen and stamens. *J. Arnold Arbor.*, **24**, 340-346.

Bailey, I. W. and Nast C. G., 1943b. The comparative morphology of the *Winteraceae*. II. Carpels. *J. Arnold Arbor.*, **24**, 472-481.

Bailey, I. W. and Nast, C. G. 1944. The comparative morphology of the *Winteraceae*. IV. Anatomy of the node and vascularization of the leaf. *J. Arnold Arbor.*, **25**, 215-221.

Bailey, I. W. and Nast, C. G. 1945a. The comparative morphology of the *Winteraceae*. VII. Summary and Conclusions. *J. Arnold Arbor.*, **26**, 37-47.

Bailey, I. W. and Nast, C. G. 1945b. Morphology and relationships of *Trochodendron* and *Tetracentron*. I. Stem, root and leaf. *J. Arnold Arbor.*, **26**, 143-154.

Bailey, I. W., Nast, C. G. and Smith, A. C. 1943. The family *Himantandraceae*. *J. Arnold Arbor.*, **24**, 190-206.

Bailey, I. W. and Smith, A. C. 1942. *Degeneriaceae*, a new family of flowering plants from Fiji. *J. Arnold Arbor.*, **23**, 356-365.

Bailey, I. W. and Swamy, B. G. L. 1948. *Amborella trichopoda* Baill. A new morphological type of vesselless dicotyledon. *J. Arnold Arbor.*, **29**, 245-254.

Bailey, I. W. and Swamy, B. G. L. 1949. The morphology and relationships of *Austrobaileya*. *J. Arnold Arbor.*, **30**, 211-226.

Bailey, I. W. and Swamy, B. G. L. 1951. The conduplicate carpel of dicotyledons and its initial trends of specialization. *Am. J. Bot.*, **38**, 373-379.

BAILEY, I. W and THOMPSON, W. P. 1918. Additional notes upon the angiosperms *Tetracentron*, *Trochodendron* and *Drimys*. *Ann. Bot.*, **37**, 503-512.

BAILLON, H. 1868-1870. Recherches organogéniques sur les *Eupomatia*. *Adansonia*, **9**, 22-28.

BAKER, H. G. 1963. Evolutionary mechanisms in pollination biology. *Science*, **139** (3558), 877-883.

BANCROFT, H. 1933. A contribution to the geographical history of the *Dipterocarpaceae*. *Geol. För. Stockh. Förh.*, **55**, 59-100.

BARANOVA, M. A. 1962. The structure of stomata and epidermal cells in Magnolias as related to the taxonomy of the genus *Magnolia* L. *Bot. Zh.*, **47**, 1108-1115. (In Russian with English summary.)

BARGHOORN, E. S. 1941. The ontogenetic development and phylogenetic specialization of rays in the xylem of dicotyledons. III. The elimination of rays. *Bull. Torrey bot. Club*, **68**, 317-325.

BARGHOORN, E. S. 1953. Climatic changes in the light of the geological past of the plant life. In *Climatic change* (Ed. H. Shapley). Cambridge.

BARLOW, B. A. 1959. Chromosome numbers in the *Casuarinaceae*. *Aust. J. Bot.*, **7**, 230-237.

BARTLING, F. G. 1830. *Ordines naturales plantarum*. Göttingen.

BAUDE, E. 1956. Die Embryoentwicklung von *Stratiotes aloides* L. *Planta*, **46**, 649-671.

BECCARI, O. 1870. Nota sull' embrione delle *Dioscoriaceae*. *Nuova G. bot. ital.*, **2**, 150.

BECK, C. B. 1962. Reconstruction of *Archaeopteris* and further consideration of its phylogenetic position. *Am. J. Bot.*, **49**, 373-382.

BECK, C. B. 1966. The origin of gymnosperms. *Taxon*, **15**, (9), 337-338.

BELL, W. A. 1956. Lower Cretaceous floras of Western Canada. *Mem. Geol. Surv. Can.*, Mem. No. 285.

BENSON, M. 1904. *Telangium scottii*, a new species of *Telangium* showing structure. *Ann. Bot.*, **18**, 161-177.

BENSON, M. S., SANDAY, E. and BERRIDGE, E. 1905. Contribution to the embryology of the *Amentiferae*. Part 2. *Carpinus betulus*. *Trans. Linn. Soc. London*, Bot., **7**, 37-44.

BENZING, D. H. 1967. Developmental patterns in stem primary xylem of woody *Ranales*. *Am. J. Bot.*, **54**, (7), 805-820.

BERG, R. L. 1956. The standardizing selection in the evolution of the flower. *Bot. Zh.*, **41**, 318-334. (In Russian.)

BERG, R. L. 1958. Further investigations in the stabilizing selection in the evolution of flower. *Bot. Zh.*, **43**, 12-27. (In Russian.)

BERRILL, N. J. 1955. *The Origin of Vertebrates.* Oxford.

BERRY, E. W. 1916. The Lower Eocene floras of southeastern North America. *Prof. Pap. U.S. geol. Surv.*, **91**.

BERRY, E. W. 1920. Palaeobotany: a sketch of the origin and evolution of floras. *Rep. Smithson Instn.*, 1918, pp. 289-407.

BERRY, E. W. 1926. Tertiary floras from British Columbia. *Bull. geol. Surv. Can.*, **42**.

BERRY, E. W. 1929. A revision of the flora of the Latah formation. *Prof. Pap. U.S. geol. Surv.*, **154**, 225-264.

BERRY, E. W. 1930a. Revision of the Lower Eocene Wilcox flora of the southeastern states. *Prof. Pap. U.S. geol. Surv.*, **156**.

BERRY, E. W. 1930b. The past climate of the north polar region. *Smithson. misc. Collns.*, **82** (6), 1-29.

BERRY, E. W. 1934. Former land connection between Asia and North America as indicated by the distribution of fossil trees. *Proc. Fifth Pacif. Sci. Congr.*, pp. 3093-3106.

BERRY, E. W. 1937. Tertiary floras of Eastern North America. *Bot. Rev.*, **3**, 31-46.

BERRY, E. W. 1945. *The Origin of Land Plants, and Four other Papers.* Baltimore.

BESSEY, C. E. 1897. Phylogeny and taxonomy of the angiosperms. *Bot. Gaz.*, **24**, 145-178.

BESSEY, C. E. 1915. The phylogenetic taxonomy of flowering plants. *Ann. Missouri bot. Gdn.*, **2**, 109-164.

BEWS, J. W. 1927. Studies in the ecological evolution of the angiosperms. *New Phytologist reprint No. 16.* London.

BHANDARI, N. N. 1963. Embryology of *Pseudowintera colorata*—a vessel-less dictotyledon. *Phytomorphology*, **13**, 303-316.

BONDESON, W. 1952. Entwicklungsgeschichte und Bau der Spaltöffnungen bei den Gattungen *Trochodendron* Sieb. et Zucc., *Tetracentron* Oliv. und *Drimys* J. R. et G. Forst. *Acta Horti Bergiani*, **16**, 169-217.

BOOTHROYD, L. E. 1930. The morphology and anatomy of the inflorescence and flower of the *Platanaceae*. *Am. J. Bot.*, **17**, 678-693.

BRAUN, A. 1864. Übersicht des natürlichen Systems. In P. ASCHERSON, *Flora der Provinz Brandenburg*, pp. 22-67. Berlin.

BRAUN, A. 1876 (1875). Die Frage nach d. *Gymnospermie d. Cycadeen*, erläutert durch die Stellung dieser Familie im Stufengang des Gewächsreichs. *Mber. dt. Akad. Wiss. Berl.*, 241-267, 289-377.

BROOKS, C. E. P. 1950. *Climate through the ages.* London.

BROWN, R. W. 1939. Fossil leaves, fruits and seeds of *Cercidiphyllum*. *J. Paleont.*, **13**, 485-499.

BROWN, R. W. 1944. Temperate species in the Eocene flora of south-eastern United States. *J. Wash. Acad. Sci.*, **34**, 349-351.

BROWN, R. W. 1946. Alterations in some fossil and living floras. *J. Wash. Acad. Sci.*, **36**, 344-355.

BROWN, R. W. 1956. Palm-like plants from the Dolores formation (Triassic) of south-western Colorado. *Prof. Pap. U.S. geol. Surv.*, **274-H**, 205-209.

BUCHHEIM, G. 1962. Beobachtungen über Bau der Frucht der *Himantandraceae*. *Sber. Ges. naturf. Freunde Berl.*, **2**, 78-92.

BUDANTSEV, L. Y. 1957. The Eocene flora of the Pavlodar Priirtyshie. In *Collection of works in memory of A. N. Kryshtofovich*, pp. 177-198. Moscow and Leningrad. (In Russian.)

BUDANTSEV, L. Y. and SVESHNIKOVA, I. N. 1964. The Tertiary flora of the Kaliningrad peninsula. IV. *Palaeobotanica*, **5**, 81-112. (In Russian with English summary.)

BURBIDGE, NANCY, T. 1960. The Phytogeography of the Australian Region. *Aust. J. Bot.*, **8**(2), 75-212.

BURCHARD, O. 1929. Beiträge zur Ökologie und Biologie der Kanarenpflanzen. *Bibl. Bot.*, Pt. 98.

CAMP, W. H. 1947. Distribution patterns in modern plants and problems of ancient dispersals. *Ecol. Monogr.*, **17**, 159-183.

CAMP, W. H. and HUBBARD, M. M. 1963. On the origin of the ovule and cupule in lyginopterid pteridosperms. *Am. J. Bot.*, **50**, 235-243.

CANRIGHT, J. E. 1952. The comparative morphology and relationships of the *Magnoliaceae*. I. Trends of specialization in the stamens. *Am. J. Bot.*, **39**, 484-497.

CANRIGHT, J. E. 1953. The comparative morphology and relationships of the *Magnoliaceae*. II. Significance of the pollen. *Phytomorphology*, **3**, 355-365.

CANRIGHT, J. E. 1955. The comparative morphology and relationships of the *Magnoliaceae*. IV. Wood and nodal anatomy. *J. Arnold Arbor.*, **36**, 119-139.

CANRIGHT, J. E. 1960. The comparative morphology and relationships of the *Magnoliaceae*. III. Carpels. *Am. J. Bot.*, **47**, 145-155.

CANRIGHT, J. E. 1963. Contributions of pollen morphology to the phylogeny of some ranalean families. *Grana palynol.*, **4**, 64-72.

CARLQUIST, S. 1961. *Comparative Plant Anatomy*. New York.

CARLQUIST, S. 1962. A theory of paedomorphosis in dicotyledonous woods. *Phytomorphology*, **12**, 30-45.

CASPARY, R. and KLEBS, R. 1907. Die Flora des Bernsteins und anderer fossiler Harze des ostpreussischen Tertiärs . . . *Abh. preuss. geol. Landesanst.*, **4**, 1-182. Berlin.

CHALK, L. 1937. The phylogenetic value of certain anatomical features of dicotyledonous woods. *Ann. Bot.*, **1**, 409-428.

CHAMBERLAIN, C. J. 1935. *Gymnosperms, structure and evolution.* Chicago.

CHANDLER, M. E. J. 1954. Some Upper Cretaceous and Eocene fruits from Egypt. *Bull. Br. Mus. nat. Hist., Geol.*, **2** (4), 147-187.

CHANDLER, M. E. J. 1958. Angiospermous fruits from the Lower Cretaceous of France and the Lower Eocene (London Clay) of Germany. *Annls. Mag. nat. Hist.* ser. 13, **1**, 354-358.

CHANDLER, M. E. J. 1961. *The Lower Tertiary Floras of Southern England.* I-II. London.

CHANDLER, M. E. J. 1964. *The Lower Tertiary Floras of Southern England.* IV. A Summary and Survey of Findings in the light of Recent Botanical Observations. London.

CHANDLER, M. E. J. and AXELROD, D. I. 1961. An early Cretaceous (Hauterivian) fruit from California. *Am. J. Sci.*, **259**, 441-446.

CHANEY, R. W. 1940. Tertiary forest and continental history. *Bull. geol. Soc. Am.*, **51**, 469-488.

CHANEY, R. W. 1944. Introduction (to Pliocene floras of California and Oregon). *Publs. Carnegie Instn.*, **553**, 1-19.

CHEADLE, V. I. 1937. Secondary growth by means of a thickening ring in certain monocotyledons. *Bot. Gaz.*, **98**, 535-555.

CHEADLE, V. I. 1953. Independent origin of vessels in the monocotyledons and dicotyledons. *Phytomorphology*, **3**, 23-44.

CHEADLE, V. I. and ESAU, K. 1958. Secondary phloem of *Calycanthaceae*. *Univ. Calif. Publs Bot.*, **29**, 397-510.

CHESTERS, K. I. M. 1955. Some plant remains from the Upper Cretaceous and Tertiary of West Africa. *Annls. Mag. nat. Hist.*, **8**, 495-504.

CHETVERIKOV, S. S. 1915. The waves of life. *Dnev. zool. Otd. imp. Obshch. Lyub. Estestvozn. Antrop. Etnogr.*, **3**, (No. 6) (*Izv. Obshch. Lyub. Estestvoz.*, **98**), pp. 106-110. St. Petersburg. (In Russian.)

CHETVERIKOV, S. S. 1961. (1926). On certain aspects of the evolutionary process from the standpoint of modern genetics. Translated from the Russian by Marina Barker. *Proc. Am. Phil. Soc.*, **105**, 167-195.

CIFERRI, R. 1962. La laurisilva canaria: una paleoflora vivente. *Ricerca scient.*, **32**, (1), 111-134.

CONWENTZ, H. 1886. *Die Flora der Bernsteins*. Vol. 2. Die Angiospermen des Bernsteins. Danzig.

CONWENTZ, H. 1890. *Monographie der baltischen Bernsteinbäume*. Danzig.

COOK, M. T. 1906. The embryology of some Cuban *Nymphaeaceae*. *Bot. Gaz.*, **42**, 376-392.

CORNER, E. J. H. 1946. Centrifugal stamens. *J. Arnold Arbor.*, **27**, 423-437.

CORNER, E. J. H. 1949a. The durian theory or the origin of the modern tree. *Ann. Bot.*, **52**, 367-414.

CORNER, E. J. H. 1949b. The annonaceous seed and its four integuments. *New Phytol.*, **48**, 332-364.

CORNER, E. J. H. 1963. A criticism of the gonophyll theory of the flower. *Phytomorphology*, **13**, 290-292.

CORNER, E. J. H. 1964. *The Life of Plants*. London.

COULTER, J. M. and CHAMBERLAIN, C. J. 1903. *Morphology of the angiosperms*. New York.

COULTER, J. M. and LAND, W. J. G. 1914. The origin of monocotyledony. *Bot. Gaz.*, **57**, 509-519.

COUPER, R. A. 1958. British Mesozoic microspores and pollen grains. *Palaeontographica*, **103** Abt. B, 75-179.

CRANWELL, LUCY, M. 1963. In J. L. GRESSITT (ed.) *Pacific Basin biogeography*, pp. 387-400. Bishop Museum Press, Honolulu.

CRÉTÉ, P. 1956. Lentibulariacées: Dévelopement de l'embryon chez *Pinguicula leptoceras* Rchb. *C. r. hebd. Séanc. Acad. Sci. Paris*, **242**, 1063-1065.

CROIZAT, L. 1940. Notes on the *Dilleniaceae* and their allies: *Austrobaileyeae*, subfam. nov. *J. Arnold Arbor.*, **21**, 397-404.

CROIZAT, L. 1947a. A study in the *Celastraceae*. *Siphonodonoideae* subfam. nov. *Lilloa*, **13**, 31-43.

CROIZAT, L. 1947b. *Trochodendron*, *Tetracentron* and their meaning in phylogeny. *Bull. Torrey bot. Club*, **74**, 60-76.

CROIZAT, L. 1952. *Manual of phytogeography*. The Hague.

CROIZAT, L. 1960. *Principia Botanica, Ia-Ib*. Codicote.

CRONQUIST, A. 1957. Outline of a new system of families and orders of dicotyledons. *Bull. Jard. bot. État Brux.*, **27**, 13-40.

CRONQUIST, A. 1960. The divisions and classes of plants. *Bot. Rev.*, **26**, 426-482.

CRONQUIST, A. 1965. The Status of the General System of Classification of Flowering Plants. *Ann. Missouri bot. Gdn.*, **52**, 281-303.

CRONQUIST, A. 1968. *The evolution and classification of flowering plants.* New York.

CRONQUIST, A., TAKHTAJAN, A. and ZIMMERMANN, W. 1966. On the higher taxa of *Embryobionta*. *Taxon*, **15**, 129-134.

CROW, L. K. 1964. The evolution of outbreeding in plants. I. The Angiosperms. *Heredity*, **19** (3), 435-457.

CZECZOTT, H. 1961. The flora of the Baltic amber and its age. *Pr. Muz. Ziemi*, **4**, 119-145. (In Polish with English summary.)

DAHL, A. O. and ROWLEY, J. R. 1965. Pollen of *Degeneria vitiensis*. *J. Arnold Arbor.*, **46**, 308-323.

DANDY, J. E. 1950. A survey of the genus *Magnolia* together with *Manglietia* and *Michelia*. *Camellias and Magnolias Conference Report*, pp. 64-81.

DANDY, J. E. and GOOD, R. D'O. 1929. *Magnoliaceae. Pflanzenareale.* 2e Reihe, Part 5, Karte, 41-43.

DANILEVSKY, A. S. and MARTYNOVA, O. M. 1962. *Lepidoptera*. In *Osnovy Paleontologii* (*Foundations of Palaeontology*), **8**, 303-307. (In Russian.)

DARLINGTON, C. D. 1963. *Chromosome botany and the origins of cultivated plants* (Revised Second Edition). London.

DARLINGTON, C. D. and MATHER, K. 1949. *The Elements of Genetics.* London.

DARLINGTON, P. J., Jr. 1957. *Zoogeography*. New York.

DARWIN, C. 1876. *The effects of cross- and self-fertilization in the vegetable kingdom*. London.

DARWIN, F. 1887. *Life and letters of Charles Darwin*. I-III.

DAUMANN, E. 1930. Das Blütennektarium von *Magnolia* und die Fütterkörper in der Blüte von *Calycanthus*. *Planta*, **11**, 108-116.

DAVIS, G. L. 1966. *Systematic embryology of the Angiosperms.* New York, London, Sydney.

DAVIS, P. H. and HEYWOOD, V. H. 1963. *Principles of Angiosperm Taxonomy*. Oliver & Boyd, Edinburgh and London.

DE BEER, G. R. 1958. *Embryos and ancestors*. Oxford.

DELEVORYAS, TH. 1962. *Morphology and evolution of fossil plants.* New York.

DELEVORYAS, TH. 1965. Investigations of North American cycadeoids: microsporangiate structures and phylogenetic implications. *Paleobotanist*, **14**, 89-93.

DELEVORYAS, TH. 1968. Investigations of North American cycadeoids: Structure, ontogeny and phylogenetic considerations of cones of *Cycadeoidea*. *Palaeontographica*, Abt. B, **121** (4-6), 122-133.

DELPINO, F. 1875. Ulteriori osservazioni sulla dicogamia nel regno vegetale. *Atti Soc. ital. Sc. nat.*, **11**, 12.

DIELS, L. 1908. *Pflanzengeographie*. Berlin.

DIELS, L. 1910. *Menispermaceae. Pflanzenreich* (ed. ENGLER), **46** (IV.94), 1-345.

DIELS, L. 1912. Über primitive Ranalen der australischen Flora. *Bot. Jb.*, **48**, Supplem. 107, 7-13.

DIELS, L. 1916. Käferblumen bei den Ranales und ihre Bedeutung für die Phylogeni der Angiospermen. *Ber. dt. bot. Ges.*, **34**, 758-774.

DIELS, L. 1919. Über die Gattung *Himantandra*, ihre Verbreitung und ihre systematische Stellung. *Bot. Jb.*, **55**, 126-134.

DIELS, L. 1932. Die Gliederung der Annonaceen und ihre Phylogenie. *Sber. preuss. Akad. Wiss.*, pp. 77-85.

DIELS, L. 1936. The genetic phytogeography of the Southwestern Pacific Area, with particular reference to Australia. In *Essays in geobotany in honor of William Albert Setchell* (ed. T. H. GOODSPEED). Berkeley, Calif.

DOBBY, E. H. 1950. *Southeast Asia*. London.

DOBZHANSKY, TH. 1951. *Genetics and the origin of species* (3rd ed.). New York and London.

DOBZHANSKY, TH. and PAVLOVSKY, O. 1957. An experimental study of interaction between genetic drift and natural selection. *Evolution*, **11**, 311-319.

DORF, E. 1960. Climatic changes of the past and present. *Am. Scient.*, **48**, 341-364.

DORMER, K. J. 1944. Some examples of correlation between stipules and lateral leaf traces. *New Phytol.*, **43**, 151-153.

DOYLE, J. and O'LEARY, M. 1934. Abnormal cones of *Fitzroya* and their bearing on the nature of the conifer strobilus. *Sci. Proc. R. Dubl. Soc.*, **21**, 23-35.

DUBININ, N. P. 1940. Darwinism and genetics of populations. *Usp. sovrem. Biol.*, **13**, 257. (In Russian.)

DUBININ, N. P. 1966. *Population Genetics*. Moscow and Leningrad. (In Russian.)

DUNN, B. D., SHARMA, G. K. and CAMPBELL, C. C. 1965. Stomatal patterns of dicotyledons and monocotyledons. *Am. Midl. Nat.*, **74**, 185-195.

DURHAM, T. W. 1963. Paleogeographic conclusions in light of biological data. In J. L. GRESSITT (ed.) *Pacific Basin biogeography*, pp. 355-365. Bishop Museum Press, Honolulu.

EAMES, A. J. 1931. The vascular anatomy of the flower with refutation of the theory of carpel polymorphism. *Am. J. Bot.*, **18**, 147-188.

EAMES, A. J. 1961. *Morphology of the angiosperms.* New York, Toronto, London.

EAMES, A. J. and MACDANIELS, L. H. 1947. *Introduction to Plant Anatomy* (2nd ed.). New York.

EARLE, T. T. 1938. Origin of the seed coats in *Magnolia. Am. J. Bot.*, **25**, 221-222.

ECKARDT, T. 1963. Some observations on the morphology and embryology of *Eucommia ulmoides. J. Indian bot. Soc.*, **42**A, 27-34.

EDWARDS, W. N. 1955. The geographical distribution of past floras. *Advmt. Sci., Lond.*, **46**, 1-12.

EHRENDORFER, F., KRENDL, F., HABELER, E. and SAUER, W. 1968. Chromosome numbers and evolution in primitive angiosperms. *Taxon*, **17**, 337-353.

EICHLER, A. W. 1878. *Blüthendiagramme.* II. Leipzig.

EL-HAMIDI, A. 1952. Vergleichend-morphologische Untersuchungen am Gynoecium der Unterfamilien *Melanthioideae* und *Asphodeloideae* der *Liliaceae. Arl. Inst. allgem. Bot. Univ. Zürich*, ser. A, **4**, 1-49.

ENDLICHER, S. 1841. *Enchiridion botanicum.* Leipzig and Vienna.

ENDRESS, P. K. 1967. Systematische Studie über die verwandtschaftlichen Beziehungen zwischen den Hamamelidaceen und Betulaceen. *Bot. Jahrb.*, **87**, (4), 431-525.

ENGLER, A. 1879, 1882. *Versuch einer Entwicklungsgeschichte der Pflanzenwelt, insbesondere der Florengebiete, seit der Tertiärperiode.* I-II. Leipzig.

ENGLER, A. 1905. Über floristische Verwandtschaft zwischen dentropischen Africa und America. *Sber. kgl. preuss. Akad. Wiss. Berl.*, 180-231.

ENGLER, A. 1914. Über Herkunft, Alter und Verbreitung extremer xerothermer Pflanzen. *Sber. kgl. preuss. Akad. Wiss.*, XX.

ERDTMAN, G. 1952. *Pollen morphology and plant taxonomy Angiosperms.* Stockholm.

ERLANSON, E. W. and HERMANN, F. J. 1928. The morphology and cytology of perfect flowers of *Populus tremuloides. Pap. Mich. Acad. Sci.*, **8**, 97-110.

ESAU, K. 1960. *Anatomy of seed plants.* New York and London.

ESAU, K. 1965. *Plant Anatomy* (2nd ed). New York.

FAEGRI, K. and VAN DER PIJL, L. 1966. *The principles of pollination ecology.* Pergamon Press, Oxford.

FAHN, A. and BAILEY, I. W. 1957. The nodal anatomy and the primary vascular cylinder of the *Calycanthaceae. J. Arnold Arbor.*, **38**, 107-117.

FEDOROV, AN. A. 1957. *Flora of Southwest China and its significance for the study of the plant world of Eurasia.* Leningrad. (In Russian.)

FEDOROV, AN. A. 1966. The structure of the tropical rain forest and speciation in the humid tropics. *J. Ecol.*, **54**, 1-11.

FLEMMING, C. A. 1963. Paleontology and southern biogeography. In J. L. GRESSITT (ed.) *Pacific Basin biogeography*, pp. 369-385. Bishop Museum Press, Honolulu.

FLORIN, R. 1931. Untersuchungen zur Stammesgeschichte der *Coniferales* und *Cordaitales*, 1. Morphologie und Epidermisstructur der Assimilationorgane bei den rezenten Coniferen. *K. Svenska Vetensk-Akad. Handl.*, **10**, 1-588.

FLORIN, R. 1933. Studien über die *Cycadales* des Mesozoikums. *K. svenska Vetensk. Akad. Handl.*, **12**, 1-134.

FLORIN, R. 1951. Evolution in cordaites and conifers. *Acta Horti Bergiani*, **15**, 285-388.

FOSTER, A. S. and GIFFORD, E. M. 1959. *Comparative morphology of vascular plants.* San Francisco.

FRIČ, A. and BAYER, E. 1903. Studie v oboru křidovêho útvaru českeho. –Perucke vrestvy. *Arch. přírodov. Výzk. Čech*, II (2). Prague.

FRIES, R. E. 1911. Ein unbeachtet gebliebenes Monokotyledonenmerkmale bei einigen *Polycarpicae. Ber. deutsch. bot. Ges.*, **29**, 292.

FRIES, R. E. 1959. *Annonaceae.* In ENGLER und PRANTL, *Die natürlichen Pflanzenfamilien* (2nd ed.), **17 a** (2), 1-171.

GARDNER, C. A. 1944. The vegetation of Western Australia. *J. Roy. Soc. W. Aust.*, **28**, 11-86.

GARDNER, J. S. 1878. How were the Eocenes of England deposited? *Pop. Sci.*, London, **2**, 282-292.

GARDNER, J. S. 1885. The Tertiary basaltic formation in Iceland. *Q. Jl. geol. Soc. Lond.*, **41**.

GARRATT, G. A. 1933. Bearing of wood anatomy on the relationships of the *Myristicaceae. Trop. Woods*, **36**, 20-44.

GARRATT, G. A. 1939. Systematic anatomy of the woods of the *Monimiaceae. Trop. Woods*, **39**, 18-44.

GARSTANG, W. 1922. The theory of recapitulation. A critical restatement of the biogenetic law. *J. Linn. Soc. Zool.*, **35**, 81-101.

GAUSSEN, H. 1946. *Les Gymnospermes, actuelles et fossilles.* Toulouse.

GILG, E. and WERDERMANN, E. 1925. *Dilleniaceae.* In ENGLER und PRANTL, *Die natürlichen Pflanzenfamilien* (2nd ed.), **21**, 7-36.

GOBI, CHR. 1916. *A review of the system of plants.* Petrograd. (In Russian with French summary.)

GOBI, CHR. 1921. Classification génétique des fruits des plantes angiospermes. *Zap. Stantsii Ispȳt. Sêmyan Imp. bot. Sada Petra Velikogo (Ann. Inst. d'essais de semences au Jard. Imper. Princ. bot. Pierre le Grand)*, **4** (4), 5-30. (In Russian with French summary.)

GOEBEL, K. 1933. *Organographie der Pflanzen*. III. Jena.

GOEPPERT, H. R. and MENGE, A. 1883, 1886. *Die Flora des Bernsteins und ihre Beziehungen zur Flora der Tertiärformation und der Gegenwart*. I-II. Danzig.

GOLENKIN, M. I. 1927. *The Victors in the struggle for existence. An investigation of the causes and conditions of the conquest of the Earth by angiosperms in the mid-Cretaceous*. Moscow. (In Russian with German summary.)

GOOD, R. 1964. *The geography of the flowering plants*. London.

GOTHAN, W. 1908. Die fossilen Hölzer von König Karls Land. *K. svenska Vetensk-Akad. Handl.*, **42** (10), 1-44.

GOTHAN, W. 1910. Die fossilen Holzreste von Spitzbergen. *K. svenska Vetensk-Akad. Handl.*, **45** (8), 1-56.

GRANT, V. 1949. Pollination systems as isolating mechanisms in angiosperms. *Evolution*, **3**, 82-97.

GRANT, V. 1950a. The protection of the ovules in flowering plants. *Evolution*, **4** (3), 179-201.

GRANT, V. 1950b. The pollination of *Calycanthus occidentalis*. *Am. J. Bot.*, **37**, 194-297.

GRANT, V. 1950c. The flower constancy of bees. *Bot. Rev.* **16**, 379-398.

GRANT, V. 1963. *The Origin of Adaptations*. New York and London.

GREGORY, M. P. 1956. A phyletic rearrangement in the *Aristolochiaceae*. *Am. J. Bot.*, **43** (2), 110-122.

GRUBOV, V. I. 1959. *Tentamen divisionis botanico-geographicae Asiae Centralis*. Leningrad. (In Russian.)

GRUSHVITSKII I. V. 1961. *The role of under-development of embryo in the evolution of flowering plants*. Moscow and Leningrad. (In Russian.)

GUNDERSEN, A. 1950. *Families of dicotyledons*. Waltham, Mass.

GUNDERSEN, A. and HASTINGS, G. T. 1944. Interdependence in plant and animal evolution. *Scient. Month. N.Y.*, **59**, 63-72.

GZYRIAN, M. S. 1952. *The family Salicaceae and its place in the system of angiosperms on the basis of the wood anatomy*. Candidate's Thesis. Erevan. (In Russian.)

HAAN, H. R. M. DE. 1920. Contribution to the knowledge of the morphological value and phylogeny of ovule and its integuments. *Recl. Trav. bot. néerl.*, **17**, 219-324.

HACCIUS, B. 1952. Die Embryoentwicklung bei *Ottelia alismoides* und das Problem des terminalen Monokotylen-Keimblattes. *Planta*, **40**, 433-460.

HACCIUS, B. 1954. Embryologische und histogenetische Studien an 'monokotylen Dikotylen'. I. *Claytonia virginica* L. *Öst. bot. Z.*, **101**, 285-303.

HALLE, T. G. 1933. The structure of certain spore-bearing organs believed to belong to the pteridosperms. *K. svenska Vetensk-Akad. Handl.*, **12** (6), 1-103.

HALLIER, H. 1901. Über die Verwandtschaftsverhältnisse der Tubifloren und Ebenalen, den polyphyletischen Ursprung der Sympetalen und Apetalen und die Anordnung der Angiospermen überhaupt. Vorstudien zum Entwurf eines Stammbaums der Blütenpflanzen. *Abh. Geb Naturw., Hamburg*, **17**, 1-112.

HALLIER, H. 1903. Vorläufiger Entwurf des natürlichen (phylogenetischen) Systems der Blütenpflanzen. *Bull. Herb. Boissier*, II, **3**, 306-317.

HALLIER, H. 1905. Provisional scheme of the natural (phylogenetic) system of flowering plants. *New Phytol.*, **4**, 151-162.

HALLIER, H. 1908. Über *Juliania*, eine Terebinthaceen-Gattung mit Cupula. *Beih. bot. Zbl.*, **23** (2), 81-265.

HALLIER, H. 1911. Über Phanerogamen von unsicherer oder unrichtiger Stellung. *Meded. Rijks-Herb.*, **1**, 1-41.

HALLIER, H. 1912a. L'origine et le système phylétique des Angiosperms exposés à l'aide de leur arbre généalogique. *Archs. néerl. Sci.*, sér. 2, **1**, 146-234.

HALLIER, H. 1912b. Über frühere Landbrücken pflanzen- und Wölkerwanderungen Zwischen Australasien und Amerika. *Med. Rijksherb., Leiden*, **13**.

HANSTEIN, J. 1870. Die Entwicklungsgeschichte der Monocotylen und Dicotylen. *Bot. Abhandl.*, Bonn, **1**, 1-112.

HARDY, A. C. 1954. Escape from specialization. In J. HUXLEY, A. C. HARDY and E. B. FORD (eds.), *Evolution as a process*, pp. 122-142. London.

HARMS, H. 1930. *Eucommiaceae*. In ENGLER and PRANTL, *Die natürlichen Pflanzenfamilien* (2nd ed.), **18a**.

HAWKES, J. G. and SMITH, P. 1965. Continental drift and the age of angiosperm genera. *Nature, Lond.*, **207** (4992), 48-50.

HAYASHI, Y. 1960. On the microsporogenesis and pollen morphology in the family *Magnoliaceae*. *Sci. Rep. Tôhoku Univ*, Ser. 4 (Biol.), **26**, 45-52.

HEER, O. 1860. *Untersuchungen über das Klima und die Vegetationsverhältnisse der Tertiärlandes*. Winterthur.

HEER, O. 1868. 1878. *Die fossile Flora der Polarländer: Flora Fossilis Arctica*. I and V. Zürich.

HEGELMAIER, F. 1874. Zur Entwicklungsgeschichte monokotyledoner Keime nebst Bemerkungen über die Bildung der Samendeckel. *Bot. Ztg.*, **39**, 631-639; **40**, 648-656; **41**, 657-671; **42**, 673-686; **43**, 689-700; **44**, 705-719.

HEGELMAIER, F. 1878. *Vergleichende Untersuchungen über Entwicklung dikotyledoner Keime mit Berücksichtigung der Pseudo-monokotyledonen*. Stuttgart.

HEMENWAY, A. F. 1913. Studies on the phloem of the dicotyledons. II, The evolution of the Sieve-tube. *Bot. Gaz.*, **55**, 236-243.

HENSLOW, G. 1888. *The origin of floral structures through insect and other agencies*. London.

HENSLOW, G. 1893. A theoretical origin of endogens from exogens through self-adaptation to an aquatic habit. *J. Linn. Soc., Bot.*, **29**, 485-528.

HENSLOW, G. 1911. The origin of monocotyledons from dicotyledons through self-adaptation to a moist or aquatic habit. *Ann. Bot.*, **26**, 717-744.

HESLOP-HARRISON, J. 1952. A reconsideration of plant teratology. *Phyton, B. Aires*, **4**, 19-34.

HESLOP-HARRISON, J. 1958a. The unisexual flower—A reply to criticism. *Phytomorphology*, **8**, 177-194.

HESLOP-HARRISON, J. 1958b. Ecological variation and ethological isolation. *Uppsala Univ. Årsskr.* **6**, 150-158.

HIEPKO, P. 1965. Vergleichend-morphologische und entwicklungsgeschichtliche Untersuchungen über das Perianth bei den *Polycarpicae*, I. Teil. *Bot. Jb.*, **84**, 359-426.

HILL, A. W. 1906. The morphology and seedling structure of the geophilous species of *Peperomia*, together with some views on the origin of monocotyledons. *Ann. Bot.*, **20**, 395-427.

HILL, A. W. 1907. A revision of the geophilous species of *Peperomia*, with some additional notes on their morphology and seedling structure. *Ann. Bot.*, **21**, 139-160.

HILL, A. W. 1920. Studies in seed germination: experiments with *Cyclamen*. *Ann. Bot.*, **34**, 417-429.

HILL, A. W. 1929. Antarctica and problems in geographical distribution. *Proc. Int. Congr. Plant Sci.*, **2**, 1474.

HILL, A. W. 1938. The monocotyledonous seedlings of certain dicotyledons, with special reference to the *Gesneriaceae*. *Ann. Bot.*, **2**, 127-143.

Hjelmqvist, H. 1948. Studies on the floral morphology and phylogeny of the *Amentiferae*. *Bot. Notiser*. Suppl., **2**, 1. Lund.

Hoffmann, E. 1948. Das Flyschproblem im Lichte der Pollenanalyse. *Phyton, Horn* (*Ann. rei bot.*), **1**, 80-101.

Holttum, R. E. 1953. Evolutionary trends in an equatorial climate. *Soc. Exper. Biol. Symposia*, **7**, 159-173.

Hooker, J. D. 1855-1860. *Flora Tasmaniae*. London.

Hooker, J. D. 1879. Presidential address to the Royal Society in 1878. *Proc. R. Soc.*, **28**, 51-55.

Hotchkiss, A. T. 1955a. Geographical distribution of the *Eupomatiaceae*. *J. Arnold Arbor.*, **36**, 385-396.

Hotchkiss, A. T. 1955b. Chromosome numbers and pollen-tetrad size in the *Winteraceae*. *Proc. Linn. Soc. N.S.W.*, **80**, 49-53.

Hotchkiss, A. T. 1958. Pollen and pollination in the *Eupomatiaceae*. *Proc. Linn. Soc. N.S.W.*, **83**, 86-91.

Hughes, N. F. and Couper, R. A. 1958. Palynology of the Brora coal of the Scottish Middle Jurassic. *Nature*, **181**, 1482-1483.

Hutchinson, J. 1923. Contributions toward a phylogenetic classification of angiosperms. II. *Annonaceae* and related forms. *Kew Bull.*, pp. 241-261.

Hutchinson, J. 1926. *The families of flowering plants*, I. London.

Hutchinson, J. 1934. *The families of flowering plants*, II. London.

Hutchinson, J. 1959a. *The families of flowering plants*, I (2nd ed.). Oxford.

Hutchinson, J. 1959b. *The families of flowering plants*, II (2nd ed.). Oxford.

Hutchinson, J. 1964. *The Genera of flowering plants. Dicotyledones.* Vol. I. Oxford.

Huxley, J. S. 1942. *Evolution, the modern synthesis*. New York.

Janaki Ammal, E. K. 1953. The Race History of the Magnolias. *Ind. J. Genet. Pl. Breed.*, **12**, 82-92.

Jeffrey, E. C. 1917. *Anatomy of woody plants*. Chicago.

Johansen, D. A. 1945. A critical survey of the present status of plant embryology. *Bot. Rev.*, **2**, 87-107.

Joshi, A. C. 1939. Morphology of *Tinospora cordifolia*, with some observations on the origin of the single integument, nature of synergidae, and affinities of *Menispermaceae*. *Am. J. Bot.*, **26**, 433-439.

Joshi, A. C. 1946. A note on the development of pollen of *Myristica fragrans* and the affinities of the family *Myristicaceae*. *J. Indian bot. Soc.*, **25**, 139-143.

JOSHI, A. C. 1947. Floral histogenesis and carpel morphology. *J. Indian bot. Soc.*, **28**, 64-74.

JUST, TH. 1948. Gymnosperms and the origin of angiosperms. *Bot. Gaz.*, **110**, 91-103.

KADEN, N. N. 1947. Genetic classification of fruits. *Vest. mosk. gos. Univ.* (*Moscow Univ. Herald*), **12**, 31-42. (In Russian.)

KADEN, N. N. 1958. Fruit. *Bolsh. Sov. Encycl.*, **33**, 264-266. (In Russian.)

KENG, H. 1959. Androdioecism in the flowers of *Trochodendron aralioides*. *J. Arnold Arbor.*, **40**, 158-160.

KERNER-MARILAUN, F. 1930. *Paläoklimatologie*. Berlin.

KIMURA, Y. 1956. Système et phylogénie des Monocotylédones. *Phanérogamie* (*Notulae System.*), **15** (2), 137-159.

KIRCHHEIMER, F. 1937. Beiträge zur Kenntnis der Flora des baltischen Bernsteins I. *Beih. bot. Zbl., Abt. B*, **57**, 441-482.

KIRITCHKOVA, A. I. and BUDANTSEV, L. U. 1967. A new find of the Lower Cretaceous flora, including angiosperms, in Yakutia. *Bot. Zh.*, **52** (7), 937-943. (In Russian.)

KOLMOGOROV, A. N. 1935. Deviations from Hardy's formula under partial isolation. *Dokl. Akad. Nauk SSSR*, **3** (3), 129-132.

KOLTSOV, N. K. 1936. *The organization of the cell*. Moscow and Leningrad. (In Russian.)

KOMAR, G. A. 1965. Arils, their nature, structure, and function. *Bot. Zh.*, **50**, 715-724. (In Russian.)

KOMAROV, V. L. 1908. Introduction to the floras of China and Mongolia. *Bull. Bot. Gard. St. Petersb.*, **29**. (In Russian.)

KOZO-POLJANSKI, B. M. 1922. *An introduction to the phylogenetic systematics of the higher plants*. Voronezh. (In Russian.)

KOZO-POLJANSKI, B. M. 1928. *The ancestors of the angiosperms*. Moscow. (In Russian).

KOZO-POLJANSKI, B. M. 1937. Teratology of the flower and new questions of its theory. *Sov. Bot.*, **6**, 56-70. (In Russian.)

KOZO-POLJANSKI, B. M. 1948 (1949). To the modernization of the system of the plant world. *Trudy voronezh. gos. Univ.*, **15**, 76-129. (In Russian.)

KRASSILOV, V. A. 1965. New fruits of angiospermous plants from the Lower Cretaceous deposits of Primorie and their stratigraphic significance. *Dokl. Acad. Sci., USSR.*, **160**, 1381-1384. (In Russian.)

KRASSILOV, V. A. 1967. *The Early Cretaceous flora of Southern Primorie and its significance for stratigraphy*. Moscow. (In Russian.)

KRASSNOV, A. N. 1956. *In the tropics of Asia*. Moscow. (In Russian.)

KRÄUSEL, R. 1956. Zur Geschichte der Angiospermen. *Bot. Mag., Tokyo*, **69**, 537-543.

KRIBS, D. A. 1935. Salient lines of structural specialisation in the wood parenchyma of dicotyledons. *Bot. Gaz.*, **96**, 547-557.

KRYSHTOFOVICH, A. N. 1946. Evolution of the vegetation through the ages and its principal factors. In V. KOMAROV (ed.) *Data on the history of the flora and vegetation of the USSR*, fasc. **2**, 21-86. Moscow and Leningrad. (In Russian with English summary.)

KRYSHTOFOVICH, A. N. 1954. The origin of the xerophyllous plant formations in the light of palaeobotany. In *The deserts of the USSR and their utilization*, **2**, 583-596. Moscow and Leningrad. (In Russian.)

KRYSHTOFOVICH, A. N. 1955. Development of the phytogeographical regions of the Northern Hemisphere from the beginning of the Tertiary period. *Problems of the Geology of Asia*, **2**, 824-844. Moscow. (In Russian.)

KRYSHTOFOVICH, A. N. 1957. *Palaeobotany*. Leningrad. (In Russian.)

KRYSHTOFOVICH, A. N. 1958a. The origin of the flora of Angaraland. In V. N. SUKATSCHEV (ed.) *Materials on the history of the flora and vegetation of the USSR*, **3**, 7-41. Moscow and Leningrad. (In Russian.)

KRYSHTOFOVICH, A. N. 1958b. The fossil floras of Penjin Bay, Lake Tastakh and Rarytkin Range. *Trudȳ bot. Inst. Akad. Nauk SSSR, Ser.* 8 (Palaeobotany), **3**, 73-121. (In Russian with English summary.)

KRYSHTOFOVICH, A. N. and PRYNADA, V. D. 1932. Data on the Mesozoic flora of the Ussuriland. *Bull. of the United Geol. and Prospecting Service of the USSR*, **22**, 1-11. (In Russian.)

KUGLER, H. 1955. *Einführung in die Blütenbiologie*. Jena.

KUHN, O. 1955. Das erste Dicotylenblatt aus dem Jura. *Orion*, 10 Jahrgang, N⁰ 19-20, 802-803.

KUPRIANOVA, L. A. 1965. *The Palynology of the Amentiferae*. Moscow and Leningrad. (In Russian.)

KUZNETSOV, N. I. 1936. *An introduction to the systematics of flowering plants*. Leningrad. (In Russian.)

LADD, H. S. 1960. Origin of the Pacific island molluscan fauna. *Am. J. Sci.*, **258**A, 137-150.

LAM, H. J. 1961. Reflections on angiosperm phylogeny—facts and theories. *Proc. K. Akad. Wet.*, Section C, **64**, 251-276.

LAVRENKO, E. M. 1962. *Salient features of the botanical Geography of the Deserts of Eurasia and Northern Africa*. Moscow and Leningrad.

LAWTON, J. R. S. and LAWTON, JUNE T. 1967. The morphology of the dormant embryo and young seedling of five species of *Dioscorea* from Nigeria. *Proc. Linn. Soc. London*, **178** (2), 153-159.

LEMESLE, R. 1938. Contribution à l'étude de l'*Eupomatia* R. Br. *Rev. gén. Bot.*, **50**, 692-712.

LEMESLE, R. 1955. Contribution à l'étude de quelques familles de dicotylédones considérées comme primitives. *Phytomorphology*, **5**, 11-45.

LEMESLE, R. and DUCHAIGNE, A. 1955. Contribution à l'étude histologique et phylogénétique du *Degeneria vitiensis* I. W. Bailey et A. C. Smith. *Rev. gén. Bot.*, **62**, 1-12.

LEMESLE, R. and PICHARD, Y. 1954. Les caractères histologiques du bois des Monimacées. *Rev. gén. Bot.*, **61**, 69-96.

LEPPIK, E. E. 1957. Evolutionary relationship between entomophilous plants and anthophilous insects. *Evolution*, **11**, 466-481.

LEPPIK, E. E. 1963. Fossil evidence of floral evolution. *Lloydia*, **26**, 91-115.

LEVINA, R. E. 1957. *Means of fruit and seed dispersal*. Moscow. (In Russian.)

LEVINA, R. E. 1967. *Fruits*. Saratov. (In Russian.)

LEWITSKY, G. A. 1931. Chromosome morphology and the Karyotype in Systematics. *Trudy̆ prikl. Bot. Genet. Selek.*, **27** (1), 187-239. (In Russian and English.)

LI, H. L. 1953. Endemism in the ligneous flora of Eastern Asia. *Proc. Seventh Pacif. Sci. Congr.*, **5**, 212-216.

LI, H. M. 1965. Plant remains of the Palaeogene age from the province Hunan. *Acta palaeontol. Sinica*, **13** (3), 540-547. (In Chinese and Russian.)

LIGNIER, O. 1908. Végétaux fossiles de Normandie. V. Nouvelles recherches sur le *Propalmophyllum liasinum* Lignier. *Mém. Soc. linn. Normandie*, **23**, 1-14.

LLOYD, F. E. 1942. *Carnivorous plants*. Waltham, U.S.A.

LONG, A. G. 1960. On the structure of *Calymmatotheca kidstonii* Calder (emended) and *Genomosperma latens* gen. and sp. nov. from the Calciferous Sandstone series of Berwickshire. *Trans. R. Soc. Edinb.*, **64**, 29-44.

LUBIMENKO, V. N. 1933. A contribution to the theory of the process of adaptation in the plant world. *Priroda* (Nature), **5-6**, 42-53; **3**, 23-35. (In Russian.)

MCLAUGHLIN, E. 1933. Systematic anatomy of the woods of the *Magnoliales*. *Trop. Woods*, **34**, 3-38.

MAHESHWARI, P. 1950. *An introduction to the embryology of angiosperms.* New York.

MALYSHEV, S. I. 1964. The role of flowering plants in the evolution of behaviour in the wasp-like ancestors of the apoids (*Vespiformia* s. lat.). *Advances in Contemporary Biology*, **57**, 159-174. (In Russian.)

MANNING, A. 1957. Some evolutionary aspects of the flower constancy of bees. *Proc. Roy. Phys. Soc. Edinb.*, **25**, 67-71.

MANUM, S. 1962. *Studies in the Tertiary flora of Spitzbergen, with notes on Tertiary floras of Elesmere Island, Greenland and Iceland.* A palynological investigation. Oslo, Norsk polarinst.

MARSDEN, M. P. F. and BAILEY, I. W. 1955. A fourth type of nodal anatomy in dicotyledons, illustrated by *Clerodendron trichotomum* Thunb. *J. Arnold Arbor.*, **36**, 1-50.

MARTIN, A. C. 1946. The comparative internal morphology of seeds. *Am. Midl. Nat.*, **36**, 513-660.

MATHER, K. 1947. Species crosses in *Antirrhinum*. I. Genetic isolation of the species *majus*, *glutinosum*, and *orontium*. *Heredity*, **1**, 175-186.

MATHEW, W. D. 1915. Climate and evolution. *Ann. New York Acad. Sc.*, :**24** 171-218; reprinted in 1939 as *Special Publ. New York Acad. Sci.*, **1**, 1-223.

MAYR, E. 1954. Change of genetic environment and evolution. In J. HUXLEY, A. C. HARDY and E. B. FORD (eds.) *Evolution as a process*, pp. 157-180. London.

MAYR, E. 1963. *Animal species and evolution*. Harvard University Press.

MEEUSE, A. D. J. 1963. From ovule to ovary: a contribution to the phylogeny of the megasporangium. *Acta biotheor.*, **16**, 127-182.

MEEUSE, A. D. J. 1965, Angiosperms, past and present. *Advg. Front. Pl. Sci.*, **11**, 1-211.

MELVILLE, R. 1962. A New Theory of the Angiosperm Flower. I. The Gynoecium. *Kew Bull.*, **16**, 1-50.

MELVILLE, R. 1963. A New Theory of the Angiosperm Flower. II. The Androecium., *Kew Bull.*, **17**, 1-63.

MENARD, H. W. 1964. *Marine Geology of the Pacific*. New York.

MENARD, H. W. and HAMILTON, E. L. 1963. Paleogeography of the tropical Pacific. In J. L. GRESSITT (ed.) *Pacific Basin biogeography*, pp. 193-217. Bishop Museum Press, Honolulu.

MERRILL, E. D. 1946. *Plant life of the Pacific world*. New York.

METCALFE, C. R. 1936. An interpretation of the morphology of the single cotyledon of *Ranunculus ficaria* based on embryology and seedling anatomy. *Ann. Bot.*, **50**, 103-120.

METCALFE, C. R. and CHALK, L. 1950. *Anatomy of the dicotyledons.* I. Oxford.

MEYER, K. I. 1960. On the embryology of *Nuphar luteum* Sm. *Byull. mosk. Obshch. Ispȳt. Prir. (Bull. Mosc. Soc. Nat.)*, Biology, **65** (6), 48-58.

MEYER, N. P. 1964. Palynological researches on the family *Nymphaeaceae. Bot. Zh.*, **49**, 1422-1430. (In Russian.)

MEYER, N. P. 1966. On the development of the pollen grains of the *Helobiae* and their connexion with the *Nymphaeaceae. Bot. Zh.*, **51**, 1736-1740. (In Russian.)

MONEY, L., BAILEY, I. W. and SWAMY, B. G. L. 1950. The morphology and relationships of the *Monimiaceae. J. Arnold Arbor.*, **31**, 372-404.

MOSELEY, M. F. 1948. Comparative anatomy and phylogeny of the *Casuarinaceae. Bot. Gaz.*, **110**, 232-280.

MOSELEY, M. F., JR. 1958. Morphological studies in the *Nymphaeaceae.* I. The nature of the Stamens. *Phytomorphology*, **8**, 1-29.

MURTY, Y. S. 1959. Studies in the order *Piperales.* VII. A contribution to the study of *Saururus cernuus* L. *J. Indian bot. Soc.*, **38**, 195-203.

NÄGELI, C. VON. 1884. *Mechanisch-physiologische Theorie der Abstammungslehre.* Munich and Leipzig.

NAIR, N. C. and BAHL, P. N. 1956. Vascular anatomy of the flower of *Myristica malabarica* Lamk. *Phytomorphology*, **6**, 127-134.

NAKAI, T. 1943. Ordines, familiae, tribus, genera, sectiones, species, varietates, formae et combinationes novae a Prof. Nakai-Takenoschin adhuc ut novis edita. Appendix. Questiones characterium naturalium plantarum, etc. Tokyo.

NAKAI, T. 1949. Classes, ordines, familiae, subfamiliae, tribus, genera nova quae attinent ad plantas Koreanas. *J. Jap. Bot.*, **24**, 8-14.

NAST, C. G. 1944. The comparative morphology of the *Winteraceae.* VI. Vascular anatomy of the flowering shoot. *J. Arnold Arbor.*, **25**, 454-466.

NAST, C. G. and BAILEY, I. W. 1945. Morphology and relationships of *Trochodendron* and *Tetracentron.* II. Inflorescence, flower and fruit. *J. Arnold Arbor.*, **26**, 265-276.

NĚMEJC, F. 1956. On the problem of the origin and phylogenetic development of the angiosperms. *Sb. nár. Mus. Praze*, Sect. B **12** b (2-3), 59-143.

NETOLIZKY, F. 1926. *Anatomie der Angiospermen-Samen.* Berlin.

NICELY, K. A. 1965. A monographic study of the *Calycanthaceae. Castanea*, **30**, 38-81.

Nitzschke, J. 1914. Beiträge zur Phylogenie der Monokotylen. *Cohn's Beitr. zur Biol. Pfl.*, **7**, 223.

Novák, F. A. 1961. *Vyssi rostliny* (*Higher plants*). Praha.

Occhioni, P. 1948. Contribuição ao estudo do familia *Canellaceae*. *Archos Jard. bot., Rio de J.*, **8**, 3-165.

Oliver, D. 1891. *Eucommia ulmoides* Oliv. *Hooker's Icon. Pl.*, **10** (2), t. 1950.

Oliver, W. R. B. 1955. History of the flora of New Zealand. *Svensk bot. Tidskr.*, **49**, 9-18.

Ozenda, P. 1949. *Recherches sur les Dicotyledones apocarpiques. Contribution à l'étude des Angiospermes dites primitives.* Paris.

Ozenda, P. 1952. Remarques sur quelques interprétations de l'étamine. *Phytomorphology*, **2**, 225-231.

Paliwal, G. S. and Bhandari, N. N. 1962. Stomatal development in some *Magnoliaceae*. *Phytomorphology*, **12**, 409-412.

Pant, D. D. and Mehra, B. 1963. Development of caryophyllaceous stomata in *Asteracantha longifolia* Nees. *Ann. Bot.*, **27**, 647-657.

Pant, D. D. and Mehra, B. 1964a. Nodal anatomy in retrospect. *Phytomorphology*, **14**, 384-387.

Pant, D. D. and Mehra, B. 1964b. Ontogeny of stomata in some *Ranunculaceae*. *Flora*, **155**, 179-188.

Pant, D. D. and Gupta, K. L. 1966. Development of stomata and foliar structure of some *Magnoliaceae*. *J. Linn. Soc., Bot.*, **59**, 165-277.

Parameswaran, N. 1962. Floral morphology and embryology in some taxa of the *Canellaceae*. *Proc. Indian Acad. Sci.*, B**55** (4), 167-182.

Parkin, J. 1914. The evolution of the inflorescence. *J. Linn. Soc., Bot.*, **42**, 511-553.

Parkin, J. 1923. The strobilus theory of angiospermous descent. *Proc. Linn. Soc. Lond., Bot.*, **153**, 51-64.

Parkin, J. 1951. The protrusion of the connective beyond the anther and its bearing on the evolution of the stamen. *Phytomorphology*, **1**, 1-8.

Parkin, J. 1952. The unisexual flower—a criticism. *Phytomorphology*, **2**, 75-79.

Parkin, J. 1953. The durian theory—a criticism. *Phytomorphology*, **3**, 80-88.

Parkin, J. 1955. A plea for a simpler gynoecium. *Phytomorphology*, **5**, 46-57.

Parkin, J. 1957. The unisexual flower again—a criticism, *Phytomorphology*, **7**, 7-9.

Percival, Mary S. 1965. *Floral biology*. Pergamon Press, Oxford.

PERIASAMY, K. and SWAMY, B. G. L. 1959. Studies in the *Annonaceae*. I. Microsporogenesis in *Cananga odorata* and *Miliusa wightiana*. *Phytomorphology*, **9**, 251-263.

PERVUKHINA, N. V. 1962. On an interesting feature of the ovary of *Trochodendron*. *Bot. Zh.*, **47**, 993-995. (In Russian.)

PERVUKHINA, N. V. and YOFFE, M. D. 1962. Floral morphology of *Trochodendron* (contributions to the phylogeny of the angiosperms). *Bot. Zh.*, **47**, 1709-1730. (In Russian.)

PIJL, L. VAN DER. 1960, 1961. Ecological aspects of flower evolution. I. Phyletic evolution. II. Zoophilous flower classes. *Evolution*, **14**, 403-416; **15**, 44-59.

POKROVSKAYA, I. M. 1956. The Oligocene spore and pollen complexes of the Baltic regions of the USSR. In *Atlas of the palynological complexes of different regions of the USSR*, pp. 7-8. Moscow. (In Russian.)

POKROVSKAYA, I. M. and STELMAK, N. K. 1960. Atlas of the Upper Cretaceous, Palaeocene and Eocene spore-pollen complexes of some regions of the USSR. *Trudȳ vses. neft. nauchno-issled. geol.-razv. Inst.* (All-Union Research Inst. of Geology), new ser., **30**, 1-575. (In Russian.)

POKROVSKAYA, I. M. and ZAUER, V. V. 1960a. The Eocene and Lower Oligocene spore and pollen complexes of the Baltic regions. In *Atlas of the Upper Cretaceous, Palaeocene and Eocene palynological complexes of some regions of the USSR*, pp. 70-81. Leningrad. (In Russian.)

POKROVSKAYA, I. M. and ZAUER, V. V. 1960b. Palynological basing of the age of the amber-bearing deposits of the Baltic region. *Dokl. Akad. Nauk SSSR*, **130**, 162-165. (In Russian.)

POPOV, M. G. 1927. Salient features of the history of development of the flora of Middle Asia. *Bull. Univ. Asia cent.* (*Byull. sredne-asiat. gos. Univ.*), **15**. Tashkent.

PORSCH, O. 1950. Geschichtliche Lebenswertung der Kastanienblüte. *Öst. bot. Z.*, **97**, 359-372.

PRANTL, K. 1887. Beiträge zur Morphologie und Systematik der Ranunculaceen. *Bot. Jb.*, **9**, 225-273.

PRITZEL, E. 1898. Der systematische Wert der Samenanatomie, insbesondere des Endosperms bei den Parietales. *Bot. Jb.*, **24**, 345-394.

RAJU, M. V. S. 1961. Morphology and anatomy of the *Saururaceae*. I. Floral anatomy and embryology. *Ann. Missouri bot. Gdn.*, **48**, 107-124.

RAO, H. S. 1939. Cuticular studies of *Magnoliales*. *Proc. Indian Acad. Sc.* B9, (2), 99-116.

Raven, P. H. and Kyhos, D. W. 1965. New evidence concerning the original basic chromosome number of Angiosperms. *Evolution*, **19**, 244-248.

Reid, E. M. and Chandler, M. E. J. 1933. *The flora of the London Clay*. London.

Remane, A. 1956. Die Grundlagen des natürlichen Systems, der vergleichenden Anatomie und der Phylogenetik (2nd ed.). *Akademische Verlagsges.*, Geest und Portig, Leipzig.

Rendle, A. B. 1938. *The classification of flowering plants*. II. Cambridge.

Richards, P. W. 1957. *The tropical rain forest. An ecological study*. Cambridge.

Robertson, C. 1904. The structure of the flowers and the mode of pollination of the primitive angiosperms. *Bot. Gaz.*, **37**, 294-298.

Rodendorf, B. B. 1962. *Insecta*. General part. In *Osnovy paleontologii (Foundations of Palaeontology)*, **8**, 29-44. (In Russian.)

Rodendorf, B. B. and Ponomarenko, A. G. 1962. *Coleoptera*. In *Foundations of palaeontology*, **9**, 241-267. (In Russian.)

Sampson, F. B. 1963. The floral morphology of *Pseudowintera*, the New Zealand member of the vesselless *Winteraceae*. *Phytomorphology*, **13**, 403-423.

Samylina, V. A. 1959. New occurrences of angiosperms from the Lower Cretaceous of the Kolyma basin. *Bot. Zh.*, **44**, 483-491. (In Russian.)

Samylina, V. A. 1960. Angiosperms from the Lower Cretaceous of the Kolyma basin. *Bot. Zh.*, **45**, 335-3 2. (In Russian with an English summary.)

Samylina, V. A. 1968. Early Cretaceous Angiosperms of the Soviet Union based on leaf and fruit remains. *J. Linn. Soc., Bot.*, **61**, 384-429.

Saporta, G. de. 1877. L'ancienne végétation polaire. *C. r. Congr. Int. Sci. Geogr., Paris.*

Saporta, G. de et Marion, A. F. 1873. Essai sur la végétation à l'époque des marnes héerisiennes de Gelinden. *Mém. cour. Mém. Sav. étr. Acad. r. Sci. Belg.*, **37**

Saporta, G. de et Marion, A. F. 1878. Revision de la flore Heerisienne de Gelinden. *Mém. cour. Mém. Sav. étr. Acad. r. Sci. Belg.*, **41**.

Saporta, G. de et Marion, A. F. 1885. *L'evolution du règne végétal. Les phanérogames.* I-II. Paris.

Sargant, E. 1903. A theory of the origin of monocotyledons, founded on the structure of their seedlings. *Ann. Bot.*, **17**, 1-92.

Sargant, E. 1904. The evolution of Monocotyledons. *Bot. Gaz.*, **37** 325-345.

SARGANT, E. 1908. The reconstruction of a race of primitive angiosperms. *Ann. Bot.*, **22**, 121-186.

SASTRI, R. L. N. 1954. On the vascular anatomy of the female flower of *Myristica fragrans*. *Proc. 41st. Indian Sci. Congr.* (Hyderabad), 132-133.

SASTRI, R. L. N. 1959. Vascularization of the carpel in *Myristica fragrans*. *Bot. Gaz.*, **121**, 92-95.

SCHAEPPI, H. 1953. Morphologische Untersuchungen an den Karpellen der *Calycanthaceae*. *Phytomorphology*, **3**, 112-120.

SCHAFFNER, J. 1904. Some morphological pecularities of the *Nymphaeaceae* and *Helobiae*. *The Ohio Naturalist*, **4**, 83-92.

SCHAFFNER, J. 1929. Principles of plant taxonomy. VII. *Ohio J. Sci.*, **29**, 243-252.

SCHAFFNER, J. 1934. Phylogenetic taxonomy of plants. *Q. Rev. Biol.*, **9**, 129-160.

SCHLOEMER-JÄGER, A. 1958. Altertiäre Pflanzen aus Flözen der Brögger-Halbinsel Spitzbergens. *Palaeontographica*, **104**B, 39-103.

SCHWARZBACH, M. 1950. *Das Klima der Vorzeit*. Stuttgart.

SCOTT, D. H. 1911. *The evolution of plants*. London and New York.

SCOTT, D. H. 1924. *Extinct plants and problems of evolution*. London.

SCOTT, R. A., BARGHOORN, E. S. and LEOPOLD, E. B. 1960. How old are the angiosperms? *Am. J. Sci.*, **258**A, 284-299.

SCOTTSBERG, C. 1925. Juan Fernandez and Hawaii. A phytogeographical discussion. *Bernice P. Bishop Museum Bulletin*, **16**, 1-47. Honolulu.

SEIDEL, C. F. 1869. Zur Entwicklungsgeschichte der *Victoria regia* Lindl. *Nov. Act. Acad. Caes. Leopoldino-Carolinae*, T.35, No. 6, 26.

SEWARD, A. C. 1910. *Fossil plants*. II. Cambridge.

SEWARD, A. C. 1933. *Plant life through the ages* (2nd ed.). Cambridge.

SEWARD, A. C. and CONWAY, V. 1935. Fossil plants from Kingigtok and Kagdlunguak, West Greenland. *Meddel. om Grønland*, **93** (5), København.

SHARP, A. J. 1951. The relation of the Eocene Wilcox flora to some modern floras. *Evolution*, **5**, 1-5.

SHIMAJI, K. 1962. Anatomical studies on the phylogenetic interrelationship of the genera in the *Fagaceae*. *Bull. Tokyo Univ. Forests*, **57**, 1-64.

SIMPSON, G. G. 1944. *Tempo and mode in evolution*. New York.

SIMPSON, G. G. 1953. *The major features of evolution*. New York.

SIMPSON, G. G. 1961. *Principles of Animal Taxonomy*. Columbia University Press, New York.

SIMPSON, J. B. 1937. Fossil pollen in Scottish Jurassic coal. *Nature*, **139**, 673.

SINNOTT, E. W. 1914. Investigations on the phylogeny of the angiosperms. 1. The anatomy of the node as an aid in the classification of angiosperms. *Am. J. Bot.*, **1**, 303-32.

SINNOT, E. W. and BAILEY, I. W. 1914. Investigations on the phylogeny of the angiosperms. 3. Nodal anatomy and the morphology of stipules. *Am. J. Bot.*, **1**, 441-453.

SKENE, MACGREGOR. 1948. *The Biology of Flowering Plants*, pp. 267-282. London.

SKVORTSOVA, N. T. 1958. On the anatomy of the flower of *Magnolia grandiflora* L. *Bot. Zh.*, **43**, 401-408. (In Russian.)

SMILEY, C. J. 1966. Cretaceous floras from Kuk River area, Alaska: stratigraphic and climatic interpretations. *Bull. geol. Soc. Am.*, **77**, 1-14.

SMITH, A. C. 1945a. Geographical distribution of the *Winteraceae*. *J. Arnold Arbor.*, **26**, 48-59.

SMITH, A. C. 1945b. A taxonomic review of *Trochodendron* and *Tetracentron*. *J. Arnold Arbor.*, **26**, 123-142.

SMITH, A. C. 1963. Summary discussion on plant distribution patterns in the tropical Pacific. In J. L. GRESSITT (ed.) *Pacific Basin biogeography*, pp. 247-249. Bishop Museum Press, Honolulu.

SMITH, A. C. 1967. The presence of primitive angiosperms in the Amazon basin and the significance in indicating migrational routes. *Atas do Symposio sobre a Biota Amazonica*, 4, (Bot.), 37-59.

SMITH, A. C. and WODEHOUSE, R. P. 1938. The American species of *Myristicaceae*. *Brittonia*, **2**, 393-510.

SMITH, D. L. 1964. The evolution of the ovule. *Biol. Rev.*, **39**, 137-159.

SNIGIREVSKAYA, N. S. 1964. Contributions to the morphology and systematics of the genus *Nelumbo* Adans. *Trudȳ bot. Inst. Akad. Nauk SSSR*, ser. I, **13**, 104-172. (In Russian.)

SOLEREDER, H. 1899. *Systematische Anatomie der Dikotyledonen*. Stuttgart.

SOLNTZEVA, M. P. and YAKOVLEV, M. S. 1964. Conditions determining the development of monocotyledony in the early embryogenesis of feather-grasses (*Stipa* spp.). *Bot. Zh.*, **49**, 625-633. (In Russian with English summary.)

SOUÈGES, R. 1954. L'origine du cône végétatif de la tige, et la question de la 'Terminalité' du cotylédon des Monocotylédones. *Anns. Sci. nat.*, XI (Bot.), **15**, 1-20.

STEBBINS, G. L. 1951. Natural selection and the differentiation of angiosperm families. *Evolution*, **5**, 299-324.

STEBBINS, G. L. 1965. The probable growth habit of the earliest flowering plants. *Ann. Mo. bot. Gdn.*, **52** (3), 457-468.

STEBBINS, G. L. 1966. Chromosomal variation and evolution. *Science*, **152**, 1463-1469.

STEBBINS, G. L. and KHUSH, G. S. 1961. Variation in the organisation of the stomatal complex in the leaf epidermis of monocotyledons, and its bearing on their phylogeny. *Am. J. Bot.*, **48**, 51-59.

STEENIS, C. G. G. J. VAN. 1951. *Flora Malesiana*, **4** (3).

STEENIS, C. G. G. J. VAN. 1962. The land-bridge theory in botany. *Blumea*, **11**, 235-372.

STEENIS, C. G. G. J. VAN. 1963. Transpacific floristic affinities, particularly in the tropical zone. In J. L. GRESSITT (ed.) *Pacific Basin biogeography*, pp. 219-231. Bishop Museum Press, Honolulu.

STERN, W. L. 1954. Xylem anatomy and relationships of *Gomortegaceae* *Am. J. Bot.*, **42**, 874-885.

STOPES, M. C. 1912. Petrifications of the earliest European angiosperms. *Phil. Trans. R. Soc.*, Ser. B, **203**, 75-100.

STOPES, M. C. 1915. The Cretaceous flora. Part II. Lower Greensand (Aptian) plants of Britain. *Catalogue Mes. Plants Brit. Mus.*, pp. 1-360.

STRAKHOV, N. M. 1960. Principles of lithogenesis. I. Moscow. (In Russian.) Eng. trans. 1966. Oliver & Boyd, Edinburgh.

SWAMY, B. G. L. 1949. Further contributions to the morphology of the *Degeneriaceae*. *J. Arnold Arbor.*, **30**, 10-38.

TAKHTAJAN, A. 1942. The structural types of gynoecium and placentation. *Bull. Armen. Branch Acad. Sc.* USSR, **3-4** (17-18), 91-112. (In Russian with English summary.)

TAKHTAJAN, A. 1943. Correlation of ontogenesis and phylogenesis in the higher plants. *Trans. Erevan State University*, **22**, 71-176. (Russian with English summary).

TAKHTAJAN, A. 1947. On principles, methods and symbols of the phylogenetic constructions in botany. *Byull. mosk. Obshch. Ispyt. Prir.*, *(Bull. Mosc. Soc. Nat.)*, Biology, **52** (5), 95-120. (In Russian.)

TAKHTAJAN, A. 1948. *Morphological evolution of the angiosperms*. Moscow. (In Russian.)

TAKHTAJAN, A. 1953. Phylogenetic principles of the system of higher plants. *Bot. Rev.*, **19**, 1-45.

TAKHTAJAN, A. 1954a. *The origin of angiospermous plants*. Moscow. (In Russian.)

TAKHTAJAN, A. 1954b. Quelques problèmes de la morphologie évolutive des angiospermes. *(Essais de Botanique)*, **2**, 763-793. Moscow and Leningrad. (In Russian and French.)

TAKHTAJAN, A. 1956. *The higher plants. 1—Psilophytales—Coniferales.* Moscow and Leningrad. (In Russian.)

TAKHTAJAN, A. 1957a. Pochodzenie okrytonasinnych (The origin of the angiosperms). *Acta Soc. Bot. Pol.*, **26**, 1-15. (In Polish and English.)

TAKHTAJAN, A. 1957b. On the origin of temperate flora of Eurasia. *Bot. Zh.*, **42**, 1635-1653. (In Russian with English summary.)

TAKHTAJAN, A. 1959. *Die Evolution der Angiospermen.* Jena.

TAKHTAJAN, A. 1961. *The origin of angiospermous plants* (2nd ed.). Moscow. (In Russian.)

TAKHTAJAN, A. 1964. *Foundations of the Evolutionary Morphology of Angiosperms.* Moscow and Leningrad. (In Russian.)

TAKHTAJAN, A. 1966. Major phytochoria of the Late Cretaceous and the Palaeocene on the territory of the USSR and adjacent countries. *Bot. Zh.*, **51**, 1217-1230. (In Russian with English summary.)

TAKHTAJAN, A. 1967. *A System and Phylogeny of the Flowering Plants.* Moscow and Leningrad. (In Russian.)

TAO, J. R. 1965. A Late Eocene florula from the district Weinan of Central Shensi. *Acta bot. Sinica*, **13** (3), 272-278. (In Chinese and English.)

TEIXEIRA, C. 1948. *Flora Mesozoica Portuguesa.* I. Lisboa.

TEIXEIRA, C. 1950. *Flora Mesozoica Portuguesa.* II. Lisboa.

TESLENKO, Y. V. 1958. Discovery of the Apt-Albian plant remains in West Siberian lowland. *Dokl. Akad. Nauk. SSSR*, **121** (5): 905-907. (In Russian.)

TESLENKO, Y. V. 1967. Some aspects of evolution of terrestrial plants. *Geologia i geofizica* (Novosibirsk), **11**, 58-64.

TESLENKO, Y. V., GOLBERT, A. V. and POLIAKOVA, I. F. 1966. The routes of dispersal of the most ancient angiosperms in Western Siberia. *Bot. Zh.*, **56** (6), 801-803. (In Russian.)

THISELTON-DYER, W. T. 1878. Lecture on plant distribution as a field for geographical research. *Geogr. J.*, **22**, 412-445.

THISELTON-DYER, W. 1909. Geographical distribution of plants. In A. C. SEWARD (ed.) *Darwin and modern science*, pp. 298-318. Cambridge.

THOMAS, H. H. 1936. Palaeobotany and the origin of the angiosperms. *Bot. Rev.*, **2**, 397-418.

THOMAS, H. H. 1947. The history of plant form. *Advmt Sci.*, 4, 243-254.

THOMPSON, W. P. 1918. Independent evolution of vessels in *Gnetales* and Angiosperms. *Bot. Gaz.*, **65**, 83-90.

THORNE, R. F. 1963. Biotic distribution patterns in the tropical Pacific. In J. L. GRESSITT (ed.) *Pacific Basin biogeography*, pp. 311-350. Bishop Museum Press, Honolulu.

THORNE, R. F. 1968. Synopsis of a putatively phylogenic classification of the flowering plants. *Aliso*, **6** (4), 57-66.

TIKHOMIROV, V. N. 1965. Some new views on the origin of flowering plants. *Trans. Moscow Soc. Naturalists*, **13**, 175-189. (In Russian with English summary.)

TIMOFEEF-RESSOVSKY, N. W. 1958. Microevolution, elementary events, material and factors of the microevolutionary process. *Bot. Zh.*, **43**, 317-336. (In Russian.)

TIPPO, O. 1938. Comparative anatomy of the *Moraceae* and their presumed allies. *Bot. Gaz.*, **100**, 1-99.

TIPPO, O. 1940. The comparative anatomy of the secondary xylem and the phylogeny of the *Eucommiaceae*. *Am. J. Bot.*, **27**, 832-838.

TOMLINSON, P. B. 1964. Stem structure in arborescent Monocotyledons. In *Formation of Wood in Forest Trees*, pp. 65-86. Academic Press, New York.

TRALAU, H. 1964. The genus *Nypa* Van Wurmb. *K. svenska Vetenskaps-Akad. Handl.*, **10**, (1), 1-29.

TRÉCUL, A. 1845. Recherches sur la structure et le développement du *Nuphar luteum*. *Ann. Sci. Nat.* sér. III. Bot. **4**, 286-345.

TUCKER, S. C. 1959. Ontogeny of the inflorescence and the flower in *Drimys winteri* var. *chilensis*. *Univ. Calif. Publs. Bot.*, **30**, 257-336.

TUCKER, S. C. and GIFFORD, E. M. 1964. Carpel vascularization of *Drimys lanceolata*. *Phytomorphology*, **14**, 197-203.

TUCKER, S. C. and GIFFORD, E. M. 1966. Organogenesis in the carpellate flower of *Drimys lanceolata*. *Am. J. Bot.*, **53**, 433-442.

UPHOF, J. C. T. 1959a. *Eupomatiaceae*. In ENGLER und PRANTL, *Die natürlichen Pflanzenfamilien* (2nd ed.), **17a**, (2), 173-176.

UPHOF, J. C. T. 1959b. *Myristicaceae*. In ENGLER und PRANTL, *Die natürlichen Pflanzenfamilien*, **17a** (2), 177-220.

UZNADZE-DGEBUADZE, M. D. 1948. The Eocene flora of the South Ural. *Proc. Geol. Inst. Acad. Sci. Georgian SSR, Tbilisi*, **4**, 161-182. (In Russian.)

VAKHRAMEEV, V. A. 1947. The role of geological surroundings in the development and dispersal of the angiospermous floras in the Cretaceous time. *Byull. mosk. Obschh. Ispÿt. Prir. Otd. Geol.* (Bulletin of the Moscow Society of Naturalists, Geology), **22** (6), 3-17. (In Russian.)

VAKHRAMEEV, V. A. 1952. The stratigraphy and fossil flora of the Cretaceous deposits of Western Kazakhstan. *Regional Stratigraphy of the USSR*, **1**. Moscow. (In Russian.)

VAKHRAMEEV, V. A. 1957. The phytogeographic and climatic zonation on the territory of Eurasia during the Jurassic and Cretaceous time. In collection of papers, *Problems in palaeobiogeography and biostratigraphy* pp. 64-76. Moscow. (In Russian.)

VAKHRAMEEV, V. A. 1966. The Late Cretaceous floras from the USSR Pacific coast, their stratigraphic range and peculiarities of composition. *Izvestia Acad. Sci. USSR*, ser. geol., 3, 76-87.

VAN DER WYK, R. W. and CANRIGHT, J. E. 1956. The anatomy and relationships of the *Annonaceae*. *Trop. Woods*, **104**, 1-24.

VAN TIEGHEM, P. 1900. Sur les dicotylédones du groups Homoxylées. *J. Bot., Paris*, **14**, 259-297, 330-361.

VAN TIEGHEM, P. and CONSTANTIN, J. 1918. *Éléments de Botanique.*

VAROSSIEAU, W. W. 1942. On the taxonomic position of *Eucommia ulmoides* Oliv. (*Eucommiaceae*). *Blumea*, **5**, 81-92.

VASILEVSKAYA, N. D. 1957. The Eocene flora of Badkhyz in Turkmenia. In *Collection of works in memory of A. N. Kryshtofovich*, pp. 103-175. Moscow and Leningrad. (In Russian.)

VAVILOV, N. I. 1926. Studies on the origin of cultivated plants. *Bull. Appl. Bot. Pl. Breed*, Leningrad, **16** (2), 1-248. (In Russian.)

VAVILOV, N. I. 1951. The origin, variation, immunity and breeding of cultivated plants. *Chron. Bot.*, **13**, 1-364.

VELENOVSKY, J. 1904. Vergleichende Studien über die *Salix*-Blüte. *Beih. bot. Zbl.*, **17** (1), 123-128.

VORONIN, N. S. 1964. Evolution of primary structure in the roots of Plants. *Scient. Papers Tsiolkovsky Pedagog. Inst. Kaluga*, **13**, 3-179. (In Russian.)

WADDINGTON, C. H. 1957. *The strategy of the genes.* London.

WAGNER, R. 1903. Beiträge zur Kenntnis der Gattung *Trochodendron* Sieb. et Zucc. *Annln naturh. Mus. Wien*, **18**, 409-422.

WALTON, J. 1953. The evolution of the ovule in the pteridosperms. *Advmt. Sci. Lond.*, **10**, 223-230.

WARDLAW, C. W. 1955. *Embryogenesis in plants.* London.

WARMING, E. 1913. Observations sur la valeur systématique de l'ovule. *Mindeskr. Japetus Steenstrups Føds.*

WEIMARCK, H. 1941. Phytogeographic groups, centres and intervals within the Cape flora. *Lunds Univ. Årsskr. N.F. Avd.* 2.Bd., **37** (5).

WHITAKER, T. W. 1933. Chromosome number and relationship in the *Magnoliales*. *J. Arnold Arbor.*, **14**, 376-385.

WHITEHOUSE, H. L. K. 1950. Multiple-allelomorph incompatibility of pollen and style in the evolution of the angiosperms. *Ann. Bot.*, **14** (54), 199-216.

WHITEHOUSE, H. L. K. 1960. Origin of Angiosperms. *Nature*, **188**, 957.

WIELAND, G. R. 1933. Origin of angiosperms. *Nature*, **131**, 360-361.

WILD, H. 1965. Additional evidence for the Africa-Madagascar-India-Ceylon land-bridge theory with special reference to the genera *Anisopappus* and *Commiphora*. *Webbia*, **19**, 497-505.

WILSON, T. K. 1960. The comparative morphology of the *Canellaceae*. I. Synopsis of genera and wood anatomy. *Trop. Woods*, **112**, 1-27.

WILSON, T. K. 1965. The comparative morphology of the *Canellaceae*. II. Anatomy of the young stem and node. *Am. J. Bot.*, **52**, 369-378.

WILSON, T. K. and MACULANS, L. M. 1967. The morphology of the *Myristicaceae*. I. Flowers of *Myristica fragrans* and *M. malabarica*. *Am. J. Bot.*, **54** (2), 214-220.

WINKLER, H. 1931. Die Monokotylen sind monokotylen. *Beitr. Biol. Pfl.*, **19**, 29-34.

WODEHOUSE, R. P. 1935. *Pollen grains*. New York.

WRIGHT, S. 1931. Evolution in Mendelian populations. *Genetics*, **16**, 97-159.

WRIGHT, S. 1948. On the roles of directed and random changes in gene frequency in the genetics of populations. *Evolution*, **2**, 279-294.

WULFF, E. V. 1943. *An Introduction to Historical Plant Geography*. Waltham, Mass.

WULFF, E. V. 1944. *Historical plant geography: history of the world flora*. Moscow and Leningrad. (In Russian.)

YAKOVLEV, M. S. 1946. Monocotyledony in the light of embryological data. *Sov. Bot.*, **14**, 351-363. (In Russian.)

YARMOLENKO, A. V. 1935. The Upper Cretaceous flora of the North-western Kara-Tau. *Acta Univ. Asiae Mediae*, Ser. 8b (Bot.), **28**, 1-36. (In Russian with English summary.)

YASUI, K. 1937. Karyological studies in *Magnolia*, with special reference to the cytokinesis in the pollen mother cell. *Bot. Mag. Tokyo*, **51**, 539-564.

YATSENKO-KHMELEVSKY, A. A. 1957. The origin of angiosperms on the data of the inner morphology of their vegetative organs. *Bot. Zh.*, **43**, 365-380. (In Russian.)

YOFFE, M. D. 1962. On the embryology of *Trochodendron aralioides* Sieb. et Zucc. (the development of pollen and embryo sac). *Trudy̆ bot. Inst. Akad. Nauk SSSR* (*Acta Inst. Bot. Acad. Sci. URSS*), ser. VII (Morphology and anatomy of plants), **5**, 250-259. (In Russian.)

YOFFE, M. D. 1965. On the embryology of *Trochodendron aralioides* Sieb. et Zucc. (Embryo and endosperm development.) In M. S. YAKOVLEV (ed.) *Flower morphology and reproductive process in angiosperms*, pp. 177-188. (In Russian.)

ZAZHURILO, K. K. 1940. On the anatomy of the seed coats of *Magnoliaceae* (*Liriodendron tulipifera* L.). *Byull. Obshch. Estest. Voronezh. gosud. Univ.* (*Bull. Soc. Nat. Voronezhsk Univ.*), 4 (1), 32-40. (In Russian.)

ZAZHURILO, K. K. and KUZNETSOVA, E. K. 1939. The nature of diffuse placentation. *Trudȳ Voronezh. gosud. Univ.* (*Acta Univ. Voroneg.*) **10** (5), 79-88. (In Russian.)

ZHILIN, S. 1968. The Oligocene flora of Ustyurt (Kazakhstan). Candidate's Thesis. Leningrad.

ZIMMERMAN, E. C. 1963. Pacific Basin biogeography. A summary discussion. In J. L. GRESSITT (ed.) *Pacific Basin biogeography*, pp. 477-481. Bishop Museum Press, Honolulu.

ZIMMERMANN, W. 1930. *Die Phylogenie der Pflanzen.* Jena.

ZIMMERMANN, W. 1959. *Die phylogenie der Pflanzen.* (2nd ed.). Stuttgart.

INDEX

A page number in **bold type** gives for a taxon its position in the author's system of classification and for a phytochorion its placing in the author's phytogeographical subdivision of the world. A number in *italics* refers to a text-figure, a Roman numeral to a plate (between p. 150 and p. 151).

N.B. The principal reference, if any, is given first, the others in orderof pagination.

N.B. To save space, generic and specific names occurring *only* in the lists on pp. 165-166 and 182-183 have *not* been indexed.